The
Maxwellians

CORNELL HISTORY OF SCIENCE SERIES

Editor: L. Pearce Williams

The Maxwellians

BRUCE J. HUNT

Cornell University Press · Ithaca and London

For my parents

First published 1991 by Cornell University Press

First printing, Cornell paperbacks, 2005

Library of Congress Cataloging-in-Publication Data

Hunt, Bruce J.
 The Maxwellians / Bruce J. Hunt.
  p. cm. — (Cornell history of science series)
 Includes bibliographical references and index.
 ISBN 978-0-8014-8234-2 (pbk. :alk. paper)
 1. Electromagnetic theory—History. 2. Physics—Great Britain—
History. 3. Maxwell, James Clerk, 1831–1879. 4. FitzGerald,
George Francis, 1851–1901. 5. Lodge, Oliver, Sir, 1851–1940.
6. Heaviside, Oliver, 1850–1925. 7. Physicists—Great Britain—
Biography. I. Title. II. Series.
QC670.H84 1991
530.1′41′094109034—dc20   91–13310

Cornell University Press strives to use environmentally responsible suppliers and materials to the fullest extent possible in the publishing of its books. Such materials include vegetable-based, low-VOC inks and acid-free papers that are recycled, totally chlorine-free, or partly composed of nonwood fibers. For further information, visit our website at www.cornellpress.cornell.edu.

Paperback printing 10 9 8 7 6 5 4 3

Contents

Illustrations

Foreword

In 1873, James Clerk Maxwell published a rambling and difficult two-volume *Treatise on Electricity and Magnetism* that was destined to change the orthodox picture of physical reality. This treatise did for electromagnetism what Newton's *Principia* had done for classical mechanics. It not only provided the mathematical tools for the investigation and representation of the whole of electromagnetic theory, but it altered the very framework of both theoretical and experimental physics. Although the process had been going on throughout the nineteenth century, it was this work that finally displaced action-at-a-distance physics and substituted the physics of the field.

Like Newton's *Principia*, Maxwell's *Treatise* did not immediately convince the scientific community. The concepts in it were strange and the mathematics was clumsy and involved. Most of the experimental basis was drawn from the researches of Michael Faraday, whose results were undeniable, but whose ideas seemed bizarre to the orthodox physicist. The British had, more or less, become accustomed to Faraday's "vision," but continental physicists, while accepting the new facts that poured from his laboratory, rejected his conceptual structures. One of Maxwell's purposes in writing his treatise was to put Faraday's ideas into the language of mathematical physics precisely so that orthodox physicists would be persuaded of their importance.

Maxwell died in 1879, midway through preparing a second edition of the *Treatise*. At that time, he had convinced only a very few of his fellow countrymen and none of his continental colleagues. That task now fell to his disciples.

The story that Bruce Hunt tells in this volume is the story of the ways in which Maxwell's ideas were picked up in Great Britain, modified, organized, and reworked mathematically so that the *Treatise* as a whole and Maxwell's concepts were clarified and made palatable, indeed irresistible, to the physicists of the late nineteenth century. The men who accomplished this, G. F. FitzGerald, Oliver Heaviside, Oliver Lodge, and others, make up the group that Hunt calls the "Maxwellians." Their relations with one another and with Maxwell's works make for a fascinating study of the ways in which new and revolutionary scientific ideas move from the periphery of scientific thought to the very center. In the process, Professor Hunt also, by extensive use of manuscript sources, examines the genesis of some of the more important ideas that fed into and led to the scientific revolution of the twentieth century.

L. PEARCE WILLIAMS

Ithaca, New York

Acknowledgments

Many people have helped to make this book possible. I am grateful for financial support given me by the National Science Foundation, the Fulbright fellowship program, Johns Hopkins University, the Smithsonian Institution, and the University of Texas. My thanks go also to the following libraries and archives for granting me access to their collections and permission to publish photographs and manuscript materials: the Royal Dublin Society; Trinity College Dublin; the Institution of Electrical Engineers; the Royal Society of London; University College London; the Deutsches Museum; the American Institute of Physics; the Syndics of the University Library, Cambridge; the Johns Hopkins University; the Air Force Geophysics Library; the University of Birmingham; and the University of Liverpool. I owe a special debt of appreciation to Professor Denis Weaire and his colleagues in the Trinity College Dublin Physics Department for their generosity in allowing me to examine and copy FitzGerald's notebooks and other papers.

I also thank my teachers, particularly Thomas Hankins, Russell McCormmach, and Robert Kargon; my former fellow students at Johns Hopkins, particularly Robert Friedman, Bob Rosenberg, and Bruce Hevly; Simon Schaffer, Andy Warwick, and the many others who made my stay in Cambridge so enjoyable; the patient and helpful editors at Cornell University Press; and my friends and colleagues at the University of Texas, especially, of course, Beth Hedrick.

BRUCE J. HUNT

Austin, Texas

References and Notation

Most of the notes refer to items in the Bibliography. Books and articles are cited by author and date; where papers are cited from collected works, the date of the original publication is given in brackets. A list of abbreviations used in the notes and Bibliography precedes the Bibliography.

I have used a rationalized vector notation based on that in Heaviside's *Electromagnetic Theory*, with two small alterations: I have used "ε" in place of Heaviside's "c" to represent the permittivity constant and "×" rather than his prefixed "V" to indicate a vector product.

Introduction

James Clerk Maxwell's theory of the electromagnetic field is generally acknowledged as one of the outstanding intellectual achievements of the nineteenth century—indeed, of any century. The late Richard Feynman once remarked, with perhaps only a little hyperbole, that "from a long view of the history of mankind . . . there can be little doubt that the most significant event of the 19th century will be judged as Maxwell's discovery of the laws of electrodynamics." Even the American Civil War, Feynman said, "will pale into provincial insignificance" beside this more profound event of the 1860s.[1] By the mid-1890s the four "Maxwell's equations" were recognized as the foundation of one of the strongest and most successful theories in all of physics; they had taken their place as companions, even rivals, to Newton's laws of mechanics. The equations were by then also being put to practical use, most dramatically in the emerging new technology of radio communications, but also in the telegraph, telephone, and electric power industries. Maxwell's theory passed to the twentieth century with an enormous reputation it has retained ever since.

It is thus perhaps surprising to find that the fullest statement Maxwell gave of his theory, his 1873 *Treatise on Electricity and Magnetism*, does not contain the four famous "Maxwell's equations," nor does it even hint at how electromagnetic waves might be produced or detected. These and many other aspects of the theory were quite thoroughly hidden in the

---

1. Feynman 1964, 2:1.11.

version of it given by Maxwell himself; in the words of Oliver Heaviside, they were "latent" in the theory, but hardly "patent."[2]

Maxwell was only forty-eight when he died of cancer in November 1879. He was only a quarter of the way through revising his *Treatise* for a second edition, and the task of digging out the "latent" aspects of his theory and of exploring its wider implications was thus left to a group of younger physicists, most of them British. Between roughly 1879 and 1894, these "Maxwellians," led by George Francis FitzGerald (1851– 1901), Oliver Lodge (1851–1940), and Oliver Heaviside (1850–1925), with a key contribution from the German physicist Heinrich Hertz (1857–1894), transformed the rich but confusing raw material of the *Treatise* into a solid, concise, and well-confirmed theory—essentially, at least for free space, the "Maxwell's theory" we know today. It was they who first explored the possibility of generating electromagnetic waves and then actually demonstrated their existence; it was they, along with J. H. Poynting (1852–1914), who first delineated the paths of energy flow in the electromagnetic field and then followed out the far-reaching implications of this discovery; it was they who recast the long list of equations Maxwell had given in his *Treatise* into the compact set now universally known as "Maxwell's"; and it was they who began to apply this revised theory to problems of electrical communications, with results that have transformed modern life. It was mainly the Maxwellians who gave Maxwell's theory the form it has since retained, and it was largely through their work that it first acquired its great reputation and breadth of application.

The evolution of "Maxwell's theory" in the years after Maxwell's death provides a striking example of a process quite common in science, as in other fields of intellectual endeavor. Scientific theories rarely spring fully formed from the mind of one person; a theory is likely to be so refined and reinterpreted by later thinkers that by the time it is codified and passes into general circulation, it often bears little resemblance to the form in which it was first propounded. The practice in science of naming theories after their originators often obscures the historical process by which scientific syntheses are achieved. One is tempted to seek all of "Newtonianism" in Newton, or all of "Darwinism" in Darwin. One of my main aims in the pages that follow is to trace the formation of such a theoretical synthesis in some detail and to show that "Maxwellianism," though undeniably built on Maxwell's ideas, was in many ways the work of his successors. "Maxwell was only ½ a Maxwellian," Heaviside declared

2. Heaviside 1892, 2:393 [1888].

in 1895; I examine here what it meant to be a Maxwellian and trace the transformation of ideas that lay behind Heaviside's remark.[3]

Another of my aims is to trace the evolution of the Maxwellians as a scientific group and to show how they stimulated and helped one another, both in their strictly scientific work and in more practical affairs. Science is a more social and cooperative process than is sometimes appreciated, and one of the most effective ways to capture its richness is to examine in detail the workings of a small group. The key to such a study of the Maxwellians is their surviving letters and notebooks, through which one can follow the course of their thoughts and actions almost day by day and see how strongly they influenced one another. In the work of FitzGerald and Lodge on ether models and electromagnetic waves; in Lodge and Heaviside's joint battles with W. H. Preece of the Post Office Telegraph Department; in Heaviside and FitzGerald's long collaboration on the problem of moving charges and on the puzzle of the ultimate nature of the electromagnetic field—in all of these, the cooperative nature of the Maxwellians' work can be clearly seen in their correspondence. Heaviside in particular virtually lived his life on paper; he was something of a recluse, and his letters and published writings were his main contact with the outside world. FitzGerald and Lodge, too, left very full records of their activities. Although all three were pioneers of electrical communications, they lived before telephones were common, and since they were physically separated—Heaviside in London and later Devon, Lodge in Liverpool, and FitzGerald in Dublin—they kept in touch mostly via letters, hundreds of which have been preserved. These enable us to reconstruct not only their work but something of their personalities and to see them engaged in the 1880s and 1890s in the lively business of remaking Maxwell's theory and of probing, as they thought, into the ultimate foundations of the physical universe.

Maxwell himself is only a minor character in this story; he died before the Maxwellians' work was well begun. But his ideas pervade the book, as they pervaded the Maxwellians' own work. Though greatly reinterpreted and recast, Maxwell's ideas always formed the core of the Maxwellian synthesis. In one of the most interesting of his unpublished writings, Heaviside reflected on the doctrine of the immortality of the soul. In its old religious sense, the idea had, he believed, been thoroughly discredited. But there was, he said, another "and far nobler sense" in which the soul truly was immortal. In living our lives, each of us "makes

---

3. Heaviside to FitzGerald, [Mar. 1895], FG-RDS; internal evidence places this undated fragment between FitzGerald's letters to Heaviside of 8 and 15 Mar., OH-IEE.

some impression on the world, good or bad, and then dies"; this impression goes on to affect future events for all time, so that "a part of us lives after us, diffused through all humanity, more or less, and all Nature. *This* is the immortality of the soul," Heaviside said. "There are large and there are small souls," he went on.

> The immortal soul of John Ploughman of Buckinghamshire is a small affair, scarcely visible. That of a Shakespeare or a Newton is stupendous. Such men live the best parts of their lives after they shuffle off the mortal coil and fall into the grave. Maxwell was one of those men. His soul will live and grow for long to come, and, thousands of years hence, it will shine as one of the bright stars of the past, whose light takes ages to reach us, amongst the crowd of others, not the least bright.[4]

This light from Maxwell has come down to us mainly through the Maxwellians; it was they who developed the most important implications of his theory and cast it into the form in which it has become most widely known. In the pages that follow, we trace how this light was refracted and refocused by the Maxwellians and how it was passed along to the next generation, to be transformed and reinterpreted again.

4. Heaviside notebook 8, OH-IEE; a slightly different version is quoted in Appleyard 1930: 257. It was probably written in 1886; cf. Heaviside 1892, 2:77 [1886].

CHAPTER ONE

FitzGerald and Maxwell's Theory

One of the last of James Clerk Maxwell's many writings on electromagnetism was a referee report he prepared for the Royal Society in February 1879. It concerned a paper by George Francis FitzGerald, and it stands as perhaps the clearest marker of the point at which Maxwell's theory passed from his own hands into those of a new generation. FitzGerald's paper, in which he extended the electromagnetic theory of light to treat reflection, refraction, and magneto-optic phenomena, was among the first to add significantly to Maxwell's original theory and had a deep and long-lasting influence on the development of electromagnetic theory. But it also reflected some of the confusions of a beginner, and Maxwell had several suggestions on how it might be improved. FitzGerald read Maxwell's remarks with great interest; although he was to become one of the leading proponents of Maxwell's electromagnetic theory, this referee report was as close as he ever came to direct contact with Maxwell himself. Even in this case, however, Maxwell's influence was in fact only posthumous: whether because of procrastination on the part of the secretary of the Royal Society, G. G. Stokes, or the onset of Maxwell's illness, the report was not forwarded to FitzGerald until 7 November 1879—two days after Maxwell's death at the early age of forty-eight.[1] FitzGerald and the other Maxwellians were left to make of Maxwell's theory what they could.

---

1. Maxwell to Stokes, 6 Feb. 1879, GGS-ULC, part pub. in Stokes 1907, 2:40–43; cf. FitzGerald 1902: 45–73 [1878].

FitzGerald and the Dublin School

FitzGerald was, by background and aptitude, unusually well equipped to take up the study of Maxwell's theory. Born 3 August 1851 in Dublin, he was a product of Trinity College Dublin and of the small but active intellectual wing of Ireland's Protestant elite. His father, William FitzGerald, was then a professor of moral philosophy at Trinity; he later served as bishop of Cork and, after 1862, of Killaloe in County Clare. He was regarded as "the most distinguished prelate in the Irish Protestant Church" and was a noted writer and metaphysician.[2] He had no flair for mathematics or science, however, and the younger FitzGerald's ability in those areas has generally been traced through his mother, Anne Stoney, and her brother George Johnstone Stoney (1826–1911), himself a noted physicist and a Fellow of the Royal Society. After the death of his mother when he was eight, FitzGerald and his brothers Maurice and William were educated at home by private tutors. Maurice (1850–1927) later taught engineering at Queen's College Belfast, while "Willy," a country clergyman at Killaloe, is best remembered as the illustrator of an early book by his Trinity College friend Bram Stoker, the author of *Dracula*.

FitzGerald's talents became obvious soon after he entered Trinity at age sixteen. According to Maurice, who also entered in 1867, his brother was "almost from the very first, first of first honours" and especially excelled at geometry: "George ran away at once in front of the class on coming to T.C.D. in that."[3] After graduating in 1871 as first senior moderator (that is, at the head of his class) in mathematics and experimental science, he settled down, "after the manner of the pick of the Dublin men," to a wide-ranging course of reading with a view to winning a fellowship.[4] Trinity College fellowships were much prized in FitzGerald's day; they offered lifetime tenure and, with time, a substantial income and considerable leisure. They fell open only when an existing Fellow died or resigned and were won by performance in a grueling four-day examination in mathematics, classics, philosophy, and the natural sciences. FitzGerald won on his second try in 1877 and began work as a college tutor; four years later he was named Erasmus Smith's Professor of Natural and Experimental Philosophy, a post he retained until his death in 1901.

---

2. Larmor 1901a, in FitzGerald 1902: xxii. The obituaries collected in FitzGerald 1902 remain the best biographical sources on FitzGerald; see also the *DNB* entries on William FitzGerald and G. J. Stoney.

3. Maurice FitzGerald to Larmor, 4 Mar. 1901, JL-RS.

4. Larmor 1901a, in FitzGerald 1902: xxii; cf. McDowell and Webb 1982: 231–32.

G. F. FitzGerald, late 1890s. From FitzGerald, *Scientific Writings*

"He had, undoubtedly, the quickest and most original brain of anybody."
　　　　　　　　—Oliver Heaviside, on hearing of FitzGerald's
　　　　　　　　　　death, February 1901

FitzGerald was tall, with a full beard even as a young man and a high forehead. He was an excellent athlete in his college days and long remained an active racquets player. His strong sense of humor often showed itself in his letters and, by all accounts, in his conversation around the Trinity dining rooms. But it was his intellectual abilities that most distinguished him, in particular his almost preternatural agility of mind. "He had, undoubtedly, the quickest and most original brain of anybody," Heaviside later said, adding that this was not always an advantage: "He saw too many openings. His brain was too fertile and inventive."[5] FitzGerald was continually tossing out new ideas but rarely followed any of them to a definite conclusion; along the way, some new possibility would occur to him and he would set off after it instead. He was not the type to sit down and quietly evolve a comprehensive theory on his own, but he excelled at drawing the best from others and was able to stimulate much good work.

The six years FitzGerald spent reading for the Trinity fellowship examinations had a deep effect on his later thought and work. It was then, for example, that he acquired his lifelong regard for the philosophy of his countryman, "the great and good Bishop Berkeley," and began to elaborate, along with his uncle G. J. Stoney, an interesting mix of philosophical idealism and practical materialism.[6] His chief study, however, was mathematical physics. He read widely in the French classics of Lagrange, Laplace, and Poisson and in some of the more recent works of the Cambridge school, including Maxwell's *Treatise*. But it was naturally in the writings of the "Dublin school" that FitzGerald most fully immersed himself and in which his own work had its deepest roots.

The mathematical traditions of Trinity College Dublin were already venerable by FitzGerald's day. Their modern flowering dated from the reforms instituted by Bartholomew Lloyd, who on becoming professor of mathematics in 1813 had set about replacing outdated English textbooks with the latest works of the French analysts and with books of his own in which the new methods were incorporated. In this reform, Dublin paralleled and even preceded Cambridge, where a similar and more famous "analytical revolution" began to bear fruit a few years later.[7] Lloyd continued his reforms after being named provost of the college, and by his death in 1837 he had made Dublin a major mathematical center. Trinity soon produced a string of distinguished mathematicians and physicists, led by William Rowan Hamilton, James MacCullagh,

5. Heaviside to John Perry, [Feb. 1901], quoted in Larmor 1901a, in FitzGerald 1902: xxvi; cf. Larmor to Heaviside, 6 Mar. 1901, OH-IEE.
6. FitzGerald 1902: 376 [1896]; see below, Chap. 4.
7. McDowell and Webb 1982: 159–61; Dubbey 1963.

Humphrey Lloyd (Bartholomew's son), J. H. Jellett, George Salmon, and FitzGerald himself. The Dublin mathematical school developed ties with the larger one at Cambridge, but it followed an independent line and kept in closer touch with Continental developments than did Cambridge. Moreover, Trinity never developed anything quite like Cambridge's Tripos examination and so escaped most of the narrow formalization that came to plague the English mathematical school by the middle of the nineteenth century.

The greatest of the Trinity mathematicians was undoubtedly Hamilton, but it was MacCullagh (1809–47), a brilliant and inspiring teacher, who did most to shape the work of the Dublin school. Known for the elegance of his geometrical methods, MacCullagh attracted a circle of devoted and talented students in both mathematics and physics. Stoney, who studied under him in the 1840s, pointed to "the close *interweaving* of analytical and geometrical conceptions and methods" as the real hallmark of MacCullagh's style, and though in the 1890s Stoney bemoaned what he saw as a decline in this tradition, an emphasis on geometrical reasoning in fact remained a cardinal feature of Dublin mathematics and physics throughout the century.[8]

The most important work of MacCullagh's short career (he committed suicide when he was thirty-eight) was in physical optics. Fresnel's wave theory of light seemed to require an elastic solid ether to support the transverse vibrations indicated by polarization phenomena, but the efforts of French physicists to devise such a medium in the 1820s and 1830s met with little success. MacCullagh made an important advance in 1839 when he showed that all of the complex phenomena of crystalline optics, including the laws of reflection, refraction, and polarization, could be derived by assuming a particular form for the Lagrangian function of the ether. In modern terms, MacCullagh made the potential energy of an element of the ether proportional to the square of its absolute rotation, or "curl," though it is not clear that he recognized this implication of his expression.[9]

While succeeding mathematical physicists acknowledged that Mac-Cullagh's equations fit the experimental evidence very well, many doubted that it was physically possible for a medium to possess the required purely rotational elasticity. Such a property was unknown in ordinary matter, and MacCullagh had proposed no plausible way in which it might be produced. Doubts about MacCullagh's rotational medium

---

8. Stoney 1897: 98; cf. McConnell 1945 and Hankins 1980: 22–24.

9. MacCullagh 1880: 145–84 [1839]; discussed and partially repr. in Schaffner 1972: 59–68, 187–93.

grew stronger after George Green's 1837 ether theory began to attract notice in the mid-1840s. Green, a Cambridge mathematician, had derived the Lagrangian function of an elastic solid in a straightforward manner and had found an expression that differed markedly from MacCullagh's. Unfortunately Green's theory also conflicted with the experimental evidence, predicting longitudinal waves that did not exist. Nonetheless, after the 1840s, most optical theorists, especially at Cambridge, preferred to build on the dynamically sound foundations Green had laid down, imposing additional conditions as needed, rather than venture on the shaky and seemingly ad hoc hypothesis MacCullagh had proposed.[10]

MacCullagh's theory suffered a further blow in 1862 when, in an influential review of optical theories, G. G. Stokes declared that it was "absolutely at variance with dynamical principles." MacCullagh's ether, he said, required "a couple . . . to act on each element to which there is no corresponding reacting couple"; it thus violated the conservation of angular momentum and so was dynamically inadmissable.[11] This objection was taken as decisive for more than thirty years, all but killing MacCullagh's theory in England. Elastic solid or "jelly" theories of Green's type, in which the ether was virtually identified with a species of ordinary matter, came to have an especially strong hold on the minds of William Thomson and G. G. Stokes, and their authority sufficed to dampen enthusiasm for alternative approaches, particularly at Cambridge, for many years.

MacCullagh's ideas survived, however, at Dublin; indeed, his papers were regarded there as classics and formed an important part of the Trinity honors curriculum. FitzGerald studied them closely while preparing for his fellowship examinations in the 1870s, and they continued to attract enough interest in 1880 to prompt two senior Trinity mathematicians, Samuel Haughton and J. H. Jellett, to assemble MacCullagh's *Collected Works* for publication by Dublin University Press.[12] (FitzGerald studied under both men and in 1885 married Jellett's daughter Harriette.) FitzGerald wrote a review of *Collected Works* in 1881 in which he expressed his admiration for MacCullagh's optical theory and his reluctance to accept Stokes's demolition of it. He said that MacCullagh had shown "the rare instinct of genius" when he proved that the complicated phenomena of crystalline optics "all flowed from a few elementary equations," and while he admitted that "doubts have been raised as to the

10. Whittaker 1951: 144–45; Schaffner 1972: 65, 93.
11. Stokes 1862; cf. Schaffner 1972: 66, 71–74.
12. MacCullagh 1880.

validity of the physical basis MacCullagh advanced for these equations," he thought they were still of great value. "Even if not well founded," FitzGerald declared, "they were well found." In particular, he argued that new developments in electromagnetic theory provided grounds for a reevaluation of MacCullagh's theory: "the most recent theories of Maxwell," he said, "while attaching new meanings to MacCullagh's symbols entirely confirm his results."[13] In his search for a sound physical basis for MacCullagh's equations, FitzGerald turned to Maxwell's field theory, a step that was to have far-reaching consequences for both theories.

Maxwell's Theory

Maxwell's theory was not entirely new when FitzGerald took it up in the 1870s; indeed, its roots went back more than forty years into the work of Michael Faraday and William Thomson (later Lord Kelvin). In the 1830s and 1840s, Faraday had begun to challenge the orthodox view that electromagnetic phenomena were the result of direct action at a distance between electrical particles and had proposed instead that they were caused by strains in an electromagnetic "field" that filled the surrounding space.[14] Faraday denied that there was any such thing as "electricity," in the sense of a substance that collected on conductors and flowed in wires. He regarded charges and currents as simply reflections of the state of the field around them, and he urged physicists to focus their attention on the stresses and strains in this field rather than on the motions of imaginary particles of "electricity." But Faraday's whole apparatus of "lines of force" and strains in the field seemed both vague and clumsy to most of his contemporaries, especially when compared with the precise and elegant action-at-a-distance theories, and field theory made little headway until Thomson took it up in the mid-1840s and began to cast it into mathematical form. Faraday's ideas were still only on the fringes of scientific respectability when, under Thomson's guidance, Maxwell began to study them in the mid-1850s.[15]

---

13. Written in the spring of 1881, this review survives in manuscript in FitzGerald notebook 10374, pp. 82–84, FG-TCD Library. It is marked "for Shaw," presumably G. F. Shaw, a Fellow of TCD and former professor of physics at Cork (McDowell and Webb 1982: 298), and was evidently intended for some Dublin or TCD periodical. Many of its remarks are echoed in a review in *Nature 24* (1881): 26–27, which is unsigned but was almost certainly written by FitzGerald.

14. Williams 1965: 202–11; Gooding 1980.

15. See Larmor 1937: 3–20.

Thomson took a major step toward elucidating the structure of the electromagnetic field in 1856 when he subjected the "Faraday effect" to dynamical analysis. Faraday had discovered in 1845 that when a polarized beam of light was sent through a piece of glass in a magnetic field, its plane of polarization was turned slightly to one side. According to Thomson, this could occur only if the magnetic field were filled with tiny "molecular vortices" spinning around the lines of force, which then passed part of their rotation to the waves of light.[16] In his 1861–62 paper "On Physical Lines of Force," Maxwell made these vortices the basis of his well-known mechanical model of the electromagnetic field. This consisted of an array of elastic vortex cells separated by layers of tiny round particles acting as "idle wheels" to transfer the spin from one vortex to the next (see Figure 1.1). The rotation of the vortices represented a magnetic field; the lateral motion of the idle-wheel particles represented an electric current. The model worked remarkably well: it gave a unified account of electromagnetic and electrostatic phenomena and could depict in considerable detail such things as the generation of magnetic fields and the induction of currents.[17] It also yielded two new ideas of great importance. First, the model suggested that during any change in the strain in an electric field, the shift of the idle-wheel particles into new positions would act as a transitory current. Such "displacement currents" were to become the keystone and most distinctive feature of Maxwell's theory. Second, Maxwell found that his elastic vortex medium would propagate waves whose velocity, calculated entirely from electromagnetic constants, was almost exactly that of light. As Maxwell exclaimed (the emphasis is his own), "We can scarcely avoid the inference that *light consists in the transverse undulations of the same medium which is the cause of electric and magnetic phenomena.*"[18]

Maxwell regarded this unification of optics and electromagnetism as perhaps his most important discovery, and he was eager to base it on something less hypothetical than his idle-wheel model. In 1864 he followed "Physical Lines" with the more abstract "Dynamical Theory of the Electromagnetic Field," in which he derived his main results, including the action of displacement currents and the electromagnetic theory of light, by a Lagrangian analysis that allowed him to skip over the detailed workings of the ether mechanism.[19] Maxwell still hoped one day to find the true mechanical structure of the ether, but until new experimental evidence allowed him to say something more definite, he thought it best

16. See the very full account in Knudsen 1976.
17. Siegel 1986 gives the best analysis of Maxwell's model.
18. Maxwell 1890, 1:500 [1862].
19. Ibid., pp. 526–97 [1864].

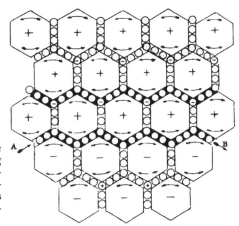

1.1. Maxwell's vortex model of the electromagnetic ether. The spinning vortices (shown as hexagons) represent a magnetic field; the lateral motion of the smaller idle wheel particles from A to B represents an electric current.

to found the laws of electromagnetism on as general and unhypothetical a basis as possible.

The papers Maxwell published in the 1860s contained all of the main points of his theory, but they had little immediate impact. It was only after his *Treatise on Electricity and Magnetism* appeared in 1873 that Maxwell's ideas began to attract much attention, and even then they spread only slowly. The *Treatise* was notoriously hard to read; it teemed with ideas but lacked the clear focus and orderly presentation that might have enabled it to win converts more readily. Rather than simply expounding his own system, Maxwell had set out to write a comprehensive treatise on electrical science, and so he had allowed his own new and distinctive ideas, notably that of the displacement current, to be almost buried under long accounts of miscellaneous phenomena discussed from several points of view. Except for a fuller treatment of the Faraday effect (in which he again invoked the molecular vortices), Maxwell added little to his earlier work on the electromagnetic theory of light; he said nothing, for example, about how electromagnetic waves might be generated, nor did he attempt to derive the laws governing reflection and refraction. Perhaps if Maxwell had lived he would, as Heaviside speculated in 1887, "in some future edition, have brought his views prominently forth *ab initio*, and developed the whole treatise on their basis exclusively."[20] But in fact Maxwell spent his last years editing the eighteenth-century manuscripts of Henry Cavendish and running the new Cavendish Laboratory at Cambridge; he was only a quarter of

20. Heaviside 1892, 2:167 [1887].

the way through revising his *Treatise* for a second edition when illness overtook him in 1879.[21] The task of extracting a cogent theory from the *Treatise* and of casting it into a form in which it could command general assent thus fell to others—to the Maxwellians, among whom FitzGerald took a leading role.

Like most of the first generation of Maxwellians, FitzGerald learned Maxwell's theory by reading the *Treatise* on his own. The copy he purchased from a Dublin bookseller in the mid-1870s is still held by the Physics Department at Trinity College Dublin, and the marginalia scattered across its pages provide an almost step-by-step picture of the evolution of FitzGerald's thinking. In them, we can see him coming to grips with the new and unfamiliar ideas, at first haltingly and then with growing confidence and facility, until at length he was able to correct and extend the theory and to apply it to new phenomena.

FitzGerald's earliest notes were those of a novice and sought mainly to clarify Maxwell's terminology and bring out the key points of the theory. He made an occasional misstep; at the end of Maxwell's discussion of self-induction, for example, FitzGerald remarked that "this opposing electromagnetic force of induction is very much like friction" and other dissipative forces that always work against motion. But he corrected this latter: "Not at all like friction," he said, "it is really quite like mass giving momentum," with the energy of induction being stored kinetically rather than dissipated.[22] As he gained a better understanding of Maxwell's theory and came to see it as a coherent whole, FitzGerald began to catch Maxwell's own slips. Thus where in his "General Equations of the Electromagnetic Field," Maxwell had expressed the force on an electric charge in terms of the scalar potential ψ, FitzGerald pointed out that this was inconsistent with the rest of the theory and that the electric force **E** should be used instead. FitzGerald published his correction in a short paper "On Electromagnetic Effects Due to the Motion of the Earth" in 1882, and it was included in the third edition of Maxwell's *Treatise*, edited by J. J. Thomson, in 1892.[23] Though seemingly a minor point—the two expressions can often be taken as equivalent—the correction had important implications for the treatment of moving charges and other questions that became prominent in the 1880s and 1890s. It also foreshadowed the general shift away from the potentials and toward the electric and magnetic force vectors that FitzGerald and especially Heaviside were later to lead.

---

21. See Chrystal 1882 on the revisions in the 2d ed.
22. See FitzGerald's copy of Maxwell 1873, 2:357, FG-TCD Physics.
23. FitzGerald 1902: 111–18 [1882]; cf. J. J. Thomson's note in Maxwell 1892, 2:258.

Reflection and Refraction

One of the longest of the notes in FitzGerald's copy of Maxwell's *Treatise* is an interleaved sheet dated 7 September 1878 in which he first broached the possibility of combining Maxwell's theory with MacCullagh's.[24] It was a step FitzGerald was peculiarly well placed to make; for he was one of the few physicists active in the 1870s who had a solid grasp of both the old and nearly forgotten ideas of MacCullagh and the new and still largely unfamiliar ones of Maxwell. The merger not only resuscitated MacCullagh's theory but extended Maxwell's in important new directions, yielding as one of its first fruits a prize that had eluded Maxwell himself: an electromagnetic theory of the reflection and refraction of light. It also raised fundamental questions about the extent to which Maxwell's theory could be reconciled with the older elastic solid theories of the ether, and thus it contributed to an important shift in the way the electromagnetic medium was conceived.

FitzGerald's search for an electromagnetic theory of reflection and refraction was prompted in part by the Glasgow physicist John Kerr's discovery in 1876 that the polarization of light was altered by reflection from the pole of a magnet.[25] FitzGerald heard of the discovery shortly after it was announced at the September meeting of the British Association, and it struck him that the theory William Thomson and Maxwell had devised to account for the Faraday effect should, if extended to reflection, explain Kerr's effect as well. The Faraday effect is essentially a double refraction for left- and right-handed circularly polarized light—that is, one has a higher index of refraction than the other and so travels more slowly through a magnetized medium. When a plane-polarized beam enters such a medium, it splits into two circularly polarized components; these get out of phase on passing through the medium and, on emerging, recombine into a beam whose plane of polarization is turned to one side, just as Faraday had observed. FitzGerald pointed out that since different indices of refraction imply different intensities of reflection, the left- and right-handed components of a polarized beam should have different amplitudes after reflection from a magnetized surface and so should recombine into an elliptically polarized beam of the kind Kerr had observed.

FitzGerald sent a short paper embodying these results to the Royal Society in November 1876, followed a week later by two supplements reporting experiments he had undertaken to test his theory. These con-

---

24. FitzGerald's copy of Maxwell 1873, sheet interleaved at 2:248, FG-TCD Physics.
25. Kerr 1876 and 1877; see also Spencer 1970.

firmed FitzGerald's predictions, and though his theory had serious deficiencies—it left aside, for example, the complications arising from metallic reflection—it gave a reasonably good qualitative explanation for Kerr's main observations and brought them into harmony with the existing theory of the Faraday effect. The paper marked a successful debut for FitzGerald, and after appearing in the *Proceedings of the Royal Society*, it was reprinted in both *Nature* and the *Philosophical Magazine*.[26]

FitzGerald spent most of 1877 preparing for his fellowship examinations and taking up his new duties as a Junior Fellow, but in 1878 he returned to the Kerr effect and began to search for a full quantitative theory of the phenomenon. In the quite natural belief that the electromagnetic theory of light offered the best prospect of accounting for the peculiarities of reflection from magnetized surfaces, he turned for guidance to Maxwell's *Treatise*. But Maxwell had in fact given no theory of the reflection of light even from ordinary surfaces, apparently because he could not settle on the proper boundary conditions, and though he treated the Faraday effect at length in his *Treatise*, he made no effort to apply his methods to magnetic reflection. "In my book I did not attempt to discuss reflexion at all," he said in his report on FitzGerald's paper in 1879. "I found that the propagation of light in a magnetized medium was a hard enough subject."[27]

FitzGerald was very familiar with MacCullagh's theory of reflection and refraction, of course, and in studying Maxwell's *Treatise,* he noticed that Maxwell's expressions for the kinetic and potential energy of the ether could be put into the same form as those used by MacCullagh. These energy expressions were very powerful tools; by applying Hamilton's principle of least action, they could be made to yield the complete behavior of a dynamical system. Simply by translating MacCullagh's equations into electromagnetic terms and then following out MacCullagh's train of analysis, FitzGerald could at a stroke extend Maxwell's theory to cover not only reflection and refraction but the whole range of crystalline optics.

In MacCullagh's theory, the kinetic energy of an element of ether is proportional to the square of its velocity, whereas the potential energy, arising solely from the rotational elasticity, is proportional to the square of the amount the element has been twisted. Mathematically, we have

26. FitzGerald 1902: 9–14 [1876]; also pub. in *Nature 15* (1877): 306–7 and *Phil. Mag. 3* (1877): 529–32.

27. Maxwell to Stokes, 6 Feb. 1879, GGS-ULC, in Stokes 1907, 2:42; cf. Maxwell to Stokes, 15 Oct. 1864, in ibid., 2:26.

$$KE_0 = \frac{1}{2} \rho \dot{\mathbf{R}}^2 \text{ (flow)},$$

$$PE_0 = \frac{1}{2} a\mathbf{T}^2 \text{ (twist)},$$

where $\mathbf{R}$ is the spatial displacement (and $\dot{\mathbf{R}}$ thus the velocity) of the ether, ρ the density, a the rotational elasticity, and $\mathbf{T}$ the amount of twist, defined by $\mathbf{T} = \text{curl } \mathbf{R}$.[28] In Maxwell's theory, as derived from his elastic vortex model, the kinetic energy of an element of free ether is proportional to the square of its rate of spin, while its potential energy is proportional to the square of the amount it has been displaced:

$$KE_1 = \frac{1}{2} \mu \mathbf{H}^2 \text{ (spin)},$$

$$PE_2 = \frac{1}{2} \epsilon \mathbf{E}^2 \text{ (displacement)},$$

where $\mathbf{H}$ is the magnetic force, $\mathbf{E}$ the electric force (and $\epsilon\mathbf{E}$ thus the displacement), and μ and ϵ the magnetic and electric constants, respectively. $\mathbf{H}$ and $\mathbf{E}$ are both related to the vector potential $\mathbf{A}$ by the equations $\mathbf{E} = -\dot{\mathbf{A}}$ and $\mu\mathbf{H} = \text{curl } \mathbf{A}$.

When FitzGerald first recognized the parallel between the two theories in September 1878, he suggested simply setting $\mathbf{R} = \mathbf{A}$.[29] This led to $\mu\mathbf{H} = \text{curl } \mathbf{R} = \mathbf{T}$ and $\mathbf{E} = -\dot{\mathbf{R}}$, and (once the constants were adjusted) made MacCullagh's energy terms equal to Maxwell's, though with the kinetic and potential terms reversed: $KE_0 = PE_1$ and $PE_0 = KE_1$. But there were problems with this approach; in particular, by making electric displacement proportional to the velocity of the ether, $\mathbf{E} = -\dot{\mathbf{R}}$, it made lines of electric force into literal lines of flow, so that ether had to be continually streaming out of positive charges and into negative ones.[30] This required a continual creation and annihilation of ether that was clearly impossible if it were to be regarded as a real substance, and FitzGerald felt compelled to abandon the idea that $\mathbf{R} = \mathbf{A}$.

28. I have somewhat simplified FitzGerald's presentation by using rationalized units and giving only the isotropic case. Both Maxwell and FitzGerald used Hamilton's quaternion notation, which is similar to the modern vector system.

29. FitzGerald's copy of Maxwell 1873, sheet interleaved at 2:248, dated "7 Sept. 1878," FG-TCD Physics.

30. The sign of this supposed flow was, of course, purely a matter of convention. Maxwell had described such a flow model, and emphasized its unreality, in his 1855–56 paper "On Faraday's Lines of Force"; see Maxwell 1890, 1:155–229.

In his paper "On the Electromagnetic Theory of the Reflection and Refraction of Light," which he sent to the Royal Society in October 1878, FitzGerald instead identified MacCullagh's **R,** the spatial displacement of the ether, with the time integral of **H,** Maxwell's magnetic force.[31] We then have $\mathbf{H} = \dot{\mathbf{R}}$, and magnetism is treated as a flow of ether. This avoids the problem of the continual creation or destruction of ether, since magnetic field lines always form closed loops. Since magnetic force and electric current are related by $\epsilon\dot{\mathbf{E}} = \text{curl } \mathbf{H}$, we have $\epsilon\dot{\mathbf{E}} = \text{curl } \dot{\mathbf{R}}$ and so $\epsilon\mathbf{E} = \text{curl } \mathbf{R} = \mathbf{T}$, so that electric "displacement" becomes a twist in the ether. MacCullagh's and Maxwell's energy terms are now identical ($\text{KE}_0 = \text{KE}_1$ and $\text{PE}_0 = \text{PE}_1$), and all of MacCullagh's results, including his theories of reflection, refraction, and double refraction, can be immediately transferred into electromagnetic terms. This accomplishment marked an enormous increase in the range and power of Maxwell's electromagnetic theory of light and made it what it had not been before: a fully fledged competitor to the existing elastic solid theories of light.

There were, however, serious obstacles to carrying through a full merger of MacCullagh's and Maxwell's theories, and these came to the fore in the second half of FitzGerald's paper, when he tried to apply the combined theory to the Kerr effect. Drawing, he said, on the vortex theory of the Faraday effect, which Maxwell had given in his *Treatise*, FitzGerald added to his expression for the kinetic energy a term intended to express the interaction of light waves and the magnetic field. But it was not clear that this magneto-optic term had a legitimate place in FitzGerald's theory, as Maxwell pointed out when the paper was sent to him to be refereed. "In order to form the equations for a magnetized medium," Maxwell wrote in his report, "Mr. FitzGerald professes to borrow an expression from my book. The expression he writes down seems to serve his purpose very well, and it may possibly be capable of reconciliation with mine, or it may possess merits altogether its own, but I have failed to trace any correspondence between them, especially in those points in which I think my own theory faulty."[32] Maxwell described his own treatment of the Faraday effect as a "hybrid" in which he had combined his electromagnetic theory of light with elements of an elastic solid theory. He had treated light waves as actual motions of the ether and had traced how these would disturb the spinning of the magnetic vortices in such a way as to cause the plane of polarization of the light to rotate. He would have preferred a purely electromagnetic theory and looked for signs of one in FitzGerald's paper. But while FitzGerald had not explicitly in-

31. FitzGerald 1902: 45–73 [1878].
32. Maxwell to Stokes, 6 Feb. 1879, GGS-ULC, in Stokes 1907, 2:42.

voked any bodily motion of the ether, it was not clear whether his argument was really independent of this assumption, or how he could obtain his results without introducing it in some fashion. It was, Maxwell said, "desirable, and even necessary to the completeness of the paper," that FitzGerald make the physical meaning of his equations more explicit; until he did so, the real nature and extent of his achievement would remain unclear.[33]

When Maxwell's remarks were forwarded to FitzGerald in November 1879, he had a hard time answering them. He had put his paper together rather haphazardly from notes, not even retaining a copy for himself, and he admitted to Stokes that he could not remember exactly what he had said. He thought he had simply copied down Maxwell's own expression for the kinetic energy in a magnetized medium, and his use of $\dot{\mathbf{R}}$ in that expression was, he said, "due to this very bodily motion [Maxwell] seems to think I have not introduced."[34] But FitzGerald had in fact not copied Maxwell's kinetic energy term exactly in his paper, nor had he made it at all clear that $\dot{\mathbf{R}}$ was to be interpreted as the velocity of the medium. He had defined $\mathbf{R}$ simply as the time integral of the magnetic force $\mathbf{H},$ and though, following MacCullagh, he evidently thought of $\mathbf{H}$ as the velocity of ether flow, he never said so in his paper. There was in fact a deep conceptual conflict between Maxwell's magneto-optic term, which was based on his vortex hypothesis and so on the identification of magnetism with *spin*, and FitzGerald's theory, in which magnetism was treated as a *flow* of ether.[35] FitzGerald's magneto-optic term was not really a consequence of the vortex hypothesis; it was, as Maxwell observed, "merely a mathematical expression from which certain observed facts may be deduced."[36]

FitzGerald responded to Maxwell's remarks by admitting, to Stokes, "My assumptions were as a matter of fact made by me from an ['analytical' deleted] mathematical point of view and not from a physical"; he had "not worked out exactly what they all come to" when interpreted physically. "This I confess is a great blot and one that I ought to have supplied," he said, adding in a characteristic bit of self-deprecation that he had not done so only because, "having got very tired of the paper long before I had finished writing it," he had been, "as I am afraid is with me

33. Ibid.; Maxwell's own treatment of the Faraday effect is in Maxwell 1873, arts. 822–30.

34. FitzGerald to Stokes, 17 Nov. 1879, GGS-ULC.

35. This is brought out in R. T. Glazebrook to FitzGerald, 5 Mar. 1881, FG-RDS, and in Glazebrook 1881.

36. Maxwell to Stokes, 6 Feb. 1879, GGS-ULC, in Stokes 1907, 2:43.

too common, too lazy to think out its consequences" and so had contented himself with just the equations.[37]

It was actually quite uncharacteristic of FitzGerald to emphasize the mathematics of a theory over its physical content. In advising FitzGerald not to worry about adding all the physical interpretations Maxwell had requested, Stokes had remarked, very acutely, "The physical bent of Maxwell's mind would naturally lead him to picture to himself a physical state, and then set himself to work out the mathematics of it." But he was well wide of the mark when he told FitzGerald, "The bent of your mind is rather to look at the mathematical expressions, and then seek for physical interpretations, or perhaps even in a great measure leave that alone."[38] While that may have been FitzGerald's procedure in this paper, his "bent of mind," as shown in almost all of his other work, was in fact intensely physical, and he had little patience with those who let the mathematics of a problem take precedence over the physics.[39] FitzGerald's usual approach was in fact very close to the one Stokes had ascribed to Maxwell.

But while FitzGerald's theory was hypothetical, it was not illegitimate. That it was not in fact based on Maxwell's molecular vortex model, as FitzGerald had claimed, was in some ways an advantage; for Maxwell had found this detour into a "hybrid" theory, in which electrical and mechanical actions were combined, the least satisfactory part of his own explanation of the Faraday effect. If FitzGerald had succeeded in avoiding such assumptions and in "explaining Kerr's phenomena as well as Faraday's by a purely electromagnetic hypothesis," his accomplishment would, Maxwell said, "be a very important step in science."[40] In a rather muddled way, FitzGerald had in fact done just that, and though his term in the kinetic energy expression was purely hypothetical—he later admitted it had "no dynamical or other foundation"—it was justified by the results that could be derived from it.[41] Moreover, it soon received support from an unexpected source: Edwin Hall's 1879 discovery that a magnetic field placed across an electric current gave rise to a small transverse voltage. Hall's professor at Johns Hopkins University, Henry

---

37. FitzGerald to Stokes, 17 Nov. 1879, GGS-ULC.
38. Stokes to FitzGerald, 30 July 1880, FG-RDS.
39. Of many examples, see FitzGerald to Rayleigh, 25 Feb. 1892, RR.11.186, RS Archives, a referee report in which FitzGerald complained that the author did not pay enough attention to the physical basis of electromagnetism in the motions of the ether; "The paper," he said, "is repulsive to me from its purely analytical character."
40. Maxwell to Stokes, 6 Feb. 1879, GGS-ULC, in Stokes 1907, 2:43.
41. FitzGerald to Rayleigh, 11 Mar. 1891, RR.11.9, RS Archives (referee report on a paper by A. B. Basset); Buchwald 1985a: 108 quotes this with the mistaken date "11 Nov. 1891."

Rowland, showed in 1880 that if an effect like the one Hall had demonstrated for conduction currents were assumed to apply to displacement currents as well—and to a Maxwellian there was every reason to think it should—it would yield a term in the electric force that would account for the Faraday effect. R. T. Glazebrook extended the analysis in 1881 and showed that the corresponding term in the kinetic energy was the same as FitzGerald's. Maxwell's slightly different term could be derived from it only by introducing the molecular vortex hypothesis. In Glazebrook's words, "Mr. Fitzgerald's term is a direct consequence of Hall's experiments; Maxwell's term is a consequence of them on some theory of the action between light and magnetism."[42] FitzGerald's work on the Kerr effect was thus an important step toward putting magneto-optics as a whole on a purely electromagnetic basis and toward uniting it with other electromagnetic phenomena.

FitzGerald's theory was far from solving all the problems of magneto-optics, and many of the deeper obscurities were not cleared up until the advent of electron theory in the mid-1890s. H. A. Lorentz and Joseph Larmor, with help from FitzGerald and others, showed that the Hall, Faraday, and Kerr effects, along with the newly discovered Zeeman effect, could all be traced to the action of the "Lorentz force," $q\mathbf{v} \times \mathbf{B}$, on electrons moving in a magnetic field.[43] This development put magneto-optics on a substantially new basis, but one that drew heavily on the earlier work of Maxwell and FitzGerald. Larmor's whole electron theory in fact grew out of his efforts in the early 1890s to explore and extend FitzGerald's electromagnetic version of MacCullagh's rotational ether, particularly in its application to magneto-optics.[44] Though seemingly a small subject, it had great strategic importance.

FitzGerald's Achievement

FitzGerald's work on reflection and refraction and on the Kerr effect strengthened Maxwell's electromagnetic theory of light in several ways. First, it demonstrated that Maxwell's theory could explain all ordinary optical phenomena at least as well as the old elastic solid theories had and that it could accommodate important new phenomena as well. Second, it

42. Glazebrook 1881: 413; Rowland 1881; cf. Buchwald 1985a: 111–17. FitzGerald remarked in a notebook entry headed "Rowland on Light" that "Glazebrook . . . has shown that Hall's (or Roland) [sic] and my terms are the same"; see FitzGerald notebook 10374, pp. 87–88, FG-TCD Library.
43. Spencer 1970: 49–51; cf. FitzGerald 1898b.
44. See below, Chap. 9, and Buchwald 1985a: 133–67.

brought out, more clearly than before, the fundamental incompatibility between Maxwell's theory and an elastic solid ether. FitzGerald had shown that Maxwell's theory was analytically equivalent to MacCullagh's, while Stokes had shown in 1862 that MacCullagh's theory, considered as an elastic solid theory, was untenable. The conclusion was inescapable: if Maxwell's theory were to survive, it had to be cut loose from reliance on an elastic solid ether and given a fundamentally new basis. Attempts to produce a "hybrid" theory, such as Maxwell had pursued in his own account of the Faraday effect, had to be abandoned. FitzGerald was convinced that slavish adherence to an elastic solid ether had been strangling optical theorizing for much too long; it had virtually killed Mac-Cullagh's theory, and now it posed a threat to Maxwell's as well. FitzGerald closed his Royal Society paper by declaring, "This investigation is put forward as a confirmation of Professor Maxwell's electromagnetic theory of light, in which, though there are some points requiring further investigation, nevertheless the foundation has certainly been laid of a very great addition to our knowledge, and if it induced us to emancipate our minds from the thraldom of a material ether might possibly lead to most important results in the theoretic explanation of nature."[45] This avowal was by no means a retreat from belief in the ether itself; FitzGerald remained wholly convinced of its existence throughout his career. Rather, it was the opening shot in his long campaign against theories of a *material* ether, in which the ether was treated as if it were a jelly or some similar kind of ordinary, elastic solid matter. FitzGerald took the lead in combating the views of Thomson and Stokes on this point throughout the 1880s and 1890s, repeatedly protesting against those who spoke of the ether "as *like* a jelly."[46] Along with the other Maxwellians, FitzGerald sought to devise a new type of ether suited to Maxwell's theory, as the elastic solid ether manifestly was not, and to put a new and less literal interpretation on Maxwell's concept of electric "displacement." He was thus one of those at the center of a major transformation in the 1880s in the prevailing view of the nature and functions of the ether and of the physical meaning of Maxwell's theory.

This transformation of ideas was closely connected with a change in personnel. It was in the late 1870s and 1880s that the Maxwellians, then in their late twenties and early thirties, began to emerge and take the lead in electromagnetic studies in Britain. Maxwell himself was dead, and most of the older physicists, particularly Thomson and Stokes,

45. FitzGerald 1902: 73 [1878].
46. FitzGerald 1902: 170–73 [1885].

proved to be too closely wedded to the old elastic solid theories to be able to assimilate the new ideas. It was left to a new generation to take up and extend the work Maxwell had begun, and by 1880, FitzGerald's papers on the electromagnetic theory of light had marked him out as one of the most promising of Maxwell's young successors.

CHAPTER TWO

FitzGerald, Lodge, and Electromagnetic Waves

That James Clerk Maxwell predicted the existence of electromagnetic waves is a commonplace of the historical literature, and Heinrich Hertz's first detection of such waves in 1888 has quite properly been hailed as a brilliant confirmation of Maxwell's theory. But the "Maxwell's theory" that Hertz's experiments confirmed was not quite the same as the one Maxwell had left at his death. Maxwell had in fact said virtually nothing about any electromagnetic waves except those of light and had given no indication of how to produce or detect relatively long waves of the kind Hertz had used. There is even some reason to think that he regarded the electrical production of such waves as an impossibility. It was only after Maxwell's death that his theory was extended to cover electromagnetic waves besides those of light, and the first unambiguous descriptions of how to generate such waves were given by FitzGerald and Lodge between 1879 and 1883. Although they did not succeed in detecting the waves themselves, their work played a crucial role in preparing the way for the reception and proper appreciation of Hertz's experiments a few years later. The extension and reinterpretation of Maxwell's theory FitzGerald and Lodge undertook was no easy task, and they made several missteps along the way. The tortuous course of their early work on electromagnetic waves illustrates very clearly how even some of the most important aspects of Maxwell's theory lay hidden at first and had to be unearthed and elaborated by the Maxwellians.

Oliver Lodge

Lodge and FitzGerald first met in August 1878 at the Dublin meeting of the British Association. Lodge had been a devoted follower of the association since 1873, when he had heard Maxwell speak at the Bradford meeting, but the 1878 meeting was apparently the first that FitzGerald attended.[1] The two young men (both were then twenty-seven) seem to have hit it off from the first and soon became close friends. Lodge said later that FitzGerald was one of the greatest influences on his life and that he "loved him as a brother," a feeling that was fully reciprocated.[2] Part of their bond was a shared enthusiasm for Maxwell's theory; they discussed it at length when together, and it was a continuing theme in their extensive correspondence. The two of them formed the core of the Maxwellian group and exerted a powerful influence on the development of British electromagnetic theory for more than twenty years.

Lodge's entry into the academic world was not as smooth as FitzGerald's. The eldest son of a self-made clay merchant, he was born in Staffordshire on 12 June 1851. His early education at a boarding school and with a clergyman uncle was old-fashioned and unpleasant; he said later, "My school days (from 8 to 14) were the dullest and most miserable that I can easily picture to myself."[3] Science formed no part of his school curriculum, but he eagerly snapped up what bits he came across on his own.

At fourteen, Lodge was taken out of school and pressed into his father's prospering business. He hated "the soul-destroying work of calling on Staffordshire potters and selling them clay," he later said, but had to spend most of seven years at it, "only squeezing out such opportunities for reading and study as I could." His enthusiasm for physics was especially fired when, at age sixteen, he spent a winter with an aunt in London and heard John Tyndall lecture at the Royal Institution. After that, he said, "I worked fiercely," taking night classes and experimenting in a small home laboratory.[4] Like many ambitious young men of the day, he set his sights on passing the entrance examinations for a University of

1. Lodge 1931b: 140 and Lodge 1931a: 45.
2. Lodge 1931b: 220; Lodge 1902, in FitzGerald 1902: xxii.
3. Lodge to John Ruskin, 7 April 1885, OJL-UB. There are several drafts of this long letter, a sort of autobiography elicited by Ruskin (whose political and economic writings Lodge greatly admired) after Lodge had written to him on some scientific points. It is an invaluable source for Lodge's view of himself in the 1880s. Much of the final version is quoted in Jolly 1974: 73–76.
4. Ibid.

London external degree. He succeeded in 1872 and, after winning a scholarship, spent the winter of 1872–73 studying at the Royal College of Science in London.

A year later Lodge broke away from his father's business for good and, after narrowly missing out on a Cambridge science scholarship, enrolled at University College London. He was an impressive figure even then, tall and a forceful and confident speaker. George Carey Foster, the professor of physics, spotted Lodge's ability and made him his assistant, enabling Lodge to support himself in a very modest way. He lived for three years in a tiny room in Camden Town (just a short way from Oliver Heaviside, though they did not meet until much later). Lodge stayed in London for several more years, taking his doctor of science degree in 1877 and holding enough miscellaneous teaching posts to enable him to marry and move into more spacious quarters. In 1881 he left to become the first professor of physics at the new University College in Liverpool, where he was to remain for nearly twenty years.

Lodge was an early convert to Maxwell's electromagnetic theory. He bought a copy of the *Treatise* after hearing it praised at the 1873 meeting of the British Association; the bookseller, he said, grumbled that it was "a product of the over-educated." But he did not make a close study of it until the summer of 1876, when on a visit to Heidelberg he spent a happy six weeks poring over the book, particularly the first volume.[5] The result was a pair of papers in the *Philosophical Magazine* describing a mechanical model he had devised to illustrate electrical phenomena on Maxwell's principles. Lodge wrote to Maxwell about this model and received a reply, "humorous and quite long," which unfortunately has not survived.[6] Lodge's mechanism of beaded strings and pulleys was not meant to represent the ether itself but, rather, to depict in an easily followed way such macroscopic phenomena as electrical conduction and dielectric polarization. It was a very effective teaching tool, and Lodge was proud of it; "I venture to think the mechanical dielectric rather nice," he told Maxwell, and he described it at length in his popular 1889 book, *Modern Views of Electricity*.[7]

Devising such models and using them to clarify and explain Maxwell's ideas was one of Lodge's greatest talents; he was a gifted expositor, with the ability to get at the conceptual core of a problem and put it in straightforward physical terms. He had no great skill in mathematical

5. Lodge 1931b: 124–25, 130.
6. Lodge 1876b, 1876a; cf. Lodge 1931b: 140.
7. Lodge to Maxwell, Nov. 1876, JCM-ULC; Lodge 1889b: 32–53. Most of *Modern Views* was first serialized in *Nature* 1887–89.

Oliver J. Lodge, 1894. From Lodge, *Past Years*

"Lodge is coming here to lecture on 'Interstellar Ether' which is I suppose the biggest thing in the Universe. Some people say that O.J.L.'s conceit of O.J.L. is a bigger, but this is an error."
—Horace Lamb to Joseph Larmor, January 1894

physics—he could follow at second hand the main arguments of others but had neither the training nor the talent to make original contributions to mathematical theory. He was a good experimenter, both in precision measurement and in more rough-and-ready exploratory work. But he was well aware of his real talents: "My strength, such as it is," he said, "lies in reasoning and brooding."[8] He especially liked to ponder what he sometimes called "the Imponderables"—light, heat, electricity, and above all the ether. It was through thinking about these, explaining them to others, and when possible, devising experiments to test his ideas, that Lodge made his main contributions to science.

In 1879, Lodge began to brood about electromagnetic waves and how, on Maxwell's theory, they might be directly produced. He was apparently the first person ever to do so. While this work did not bear immediate fruit, it influenced both his own thinking and FitzGerald's and led them into areas that Maxwell had left unexplored. By the mid-1880s they had codified as part of "Maxwell's theory" a set of ideas about electromagnetic waves that differed significantly from anything Maxwell had himself conceived.

Maxwell and Electromagnetic Waves

The basic question about Maxwell and electromagnetic waves is simple: why did he never try to produce them? Why did he never even discuss the issue with his students at the Cavendish Laboratory or in any of his electrical writings? Maxwell's silence about electromagnetic waves is well established, but the reasons for it have been a matter for some debate.

Thomas Simpson has argued that Maxwell simply was not interested in electromagnetic waves; his real interest was always in optics, not electricity, according to Simpson, and he pursued electromagnetic theory only for the sake of what it could tell him about the nature of the optical ether.[9] We should thus not be surprised that Maxwell never sought to generate or detect relatively long electromagnetic waves; his attention was always focused instead on the very short waves constituting light, which he could, of course, generate simply by striking a match. But there are several objections to Simpson's argument. Maxwell manifestly *was*

8. Lodge to Ruskin, 7 Apr. 1885, OJL-UB, quoted in Jolly 1947: 73; cf. Lodge 1931b: 111.

9. Simpson 1966, which also summarizes the strong evidence that Maxwell neither performed nor contemplated any experiments on electromagnetic waves other than those of light.

interested in electromagnetism for its own sake; he began to work on it well before he formulated this electromagnetic theory of light, and he never published much purely optical work. The theory of light occupies only a small part of his *Treatise on Electricity and Magnetism* or any of his other writings, and there is no real evidence that he studied electricity merely as a preliminary to more fundamental optical researchers. Indeed, it would seem more plausible to turn this argument on its head and to maintain that Maxwell used optical phenomena (such as the Faraday effect) to illuminate the electromagnetic questions that were his real interest.

A more satisfactory explanation of Maxwell's silence about electromagnetic waves is A. F. Chalmers's assertion that Maxwell simply never realized an oscillating current would emit electromagnetic waves.[10] Such emission is by no means an obvious consequence of his equations, especially in the form in which he originally gave them, and Chalmers makes a strong case that Maxwell missed it. This would certainly not be the only time Maxwell failed to see an important implication of his theory.

Maxwell's belief that light waves were electromagnetic does not imply that he believed they were generated electromagnetically. Instead, he seems to have regarded the production of light as a mechanical process, traceable to the vibration of the parts of molecules and thus dependent on the connection between matter and ether, something Maxwell claimed to know little about but which he did not regard as necessarily electromagnetic. Maxwell's ether was essentially mechanical; light and all electromagnetic phenomena were manifestations of the stresses and motions of that ether. He believed the laws of electromagnetism to be *consequences* of fine-grained mechanical ether motions; those laws need not themselves extend to the molecular level. This belief comes across most clearly in Maxwell's application of his molecular vortex hypothesis to the Faraday effect, in which he treated the magnetic field as an array of tiny ether vortices and light waves as mechanical displacements of the ether.[11] Since he regarded the production of light as an essentially molecular and mechanical process, prior, in a sense, to electromagnetic laws, Maxwell could elaborate an electromagnetic account of the propagation of light without ever supposing that ether waves were produced purely electromagnetically.

Thus Maxwell evidently missed what is now regarded as the most exciting implication of his theory, and one with enormous practical conse-

10. Chalmers 1974: 471–76.
11. Maxwell 1873, arts. 822–31.

quences. That relatively long electromagnetic waves, or perhaps light itself, could be generated in the laboratory with ordinary electrical apparatus was unsuspected through most of the 1870s.[12] It was only at the very end of the decade, and more thoroughly in the early 1880s, that some of Maxwell's followers began to explore this question and, after a number of false starts, to discover just how such waves could be produced.

Lodge and "Electromagnetic Light"

Late in the summer of 1879, just a few months before Maxwell's death, Lodge began to look into the possibility of producing electromagnetic waves. He recorded various ideas in his research notebooks and discussed them at the British Association meeting in August and in letters to FitzGerald the following February. It was, Lodge later said, "so far as I know . . . the first feeling after the experimental generation of electromagnetic light," and it had an important effect on later thinking about the production of electromagnetic radiation.[13]

In this work, Lodge approached Maxwell's theory not through its equations but, characteristically, through a semimechanical model he had devised to represent it. He gave a paper on the subject, "On a Hypothesis concerning the Ether in Connexion with Maxwell's Theory of Electricity," at the August 1879 meeting of the British Association; it was, he said more than twenty years later, "little more than an oral communication to Section A," and only the title appeared in the annual *Report*.[14] Our only real indication of its content comes from a letter Lodge wrote to Joseph Larmor in 1902:

> So far as I recollect it, it was to the effect that as in my electrostatic models positive and negative electricity always did opposite things—so that shear of ether explained electrostatic energy—so this oppositeness of behaviour would make them as regards rotation act as if they were geared up like cogwheels rotating in opposite directions: wherefore a disturbance would spread laterally when any part of it was twisted (magnetism); and hence that light could be got from oscillations.[15]

12. In a letter to Stokes, 15 Oct. 1864, in Stokes 1907, 2:26, Maxwell compared the velocity of light waves to that of "such slow disturbances as we can make," but it is not clear that he regarded these disturbances as actual electromagnetic *waves* of relatively long length; in any case, he said nothing about how they might be produced or detected.
13. Lodge to Larmor, 1 Jan. 1902, OJL-UCL; returned to Lodge with Larmor's marginal comments.
14. See *BA Report* (1879), p. 258.
15. Lodge to Larmor, 1 Jan. 1902, OJL-UCL.

Lodge first described this cogwheel model of the ether in print in 1888 and treated it in great detail in *Modern Views*.[16] It resembled Maxwell's well-known vortex and idle-wheel model except that Lodge's vortices were directly geared together and were of two distinct types. One kind of vortex represented positive electricity and spun in one direction; the other represented negative electricity and spun the opposite way. Lodge was able to represent most of the phenomena of electrodynamics on the model, including induction, conduction, and the propagation of waves.

It seems likely that in his 1902 recollections Lodge was projecting some of his later ideas back into his 1879 work, but he certainly had at least some of the elements of his cogwheel ether in mind in the earlier period. This is clear from a letter he wrote to FitzGerald in February 1880, in which he mentioned "my notion, which I think you don't like and which is certainly very crude, that Ether is + & − Electricity together & that when an EMF [electromotive force] is applied the ether is sheared but not moved bodily one way of the other." Lodge said he had "fancied that light . . . was periodic electrical displacements viz. + up & − down, the restoring force being due to the electrostatic strain instead of ordinary elasticity. And hence I imagined that light might be excited electrically."[17]

Thus Lodge's "very crude" model of the ether had led him to an idea that had apparently eluded Maxwell himself: that of generating waves directly by electromagnetic means. But how could he do this experimentally? Lodge recorded several ideas in his laboratory notebooks, beginning with a 3 August 1879 entry headed "Electromagnetic Light."[18] It should be stressed that Lodge was looking for *light;* he had not yet hit on the important idea, developed later by FitzGerald and Hertz, of producing and detecting relatively long electromagnetic waves. (Hertz used waves of about 3 meters, with a corresponding frequency of 10^8 cycles per second; visible light, in contrast, has a wavelength of less than 10^{-6} meters and a frequency of over 10^{14} cycles per second.) Lodge thus avoided the problem of how to detect the waves he proposed to produce—he could simply use his eyes—but he faced the daunting task of producing electrical oscillations of extraordinarily high frequency.

Lodge's early ideas were rather primitive and would not have worked. His first entry reads:

16. Lodge 1889b: 177–216; the first part first appeared in *Nature 37* (1888): 344–48. Lodge's cogwheel model was not fully mechanical, since the difference between positive and negative cogs was not explained; see below, Chap. 4.

17. Lodge to FitzGerald, 29 Feb. [1880], FG-RDS.

18. Lodge to Larmor, 1 Jan. 1902, OJL-UCL, which includes extracts from Lodge's notebooks for 1879–80, which Lodge hoped would clarify a passage in a paper of FitzGerald's that Larmor was editing; see FitzGerald 1902: 100 [1882].

> Is it possible to generate light by magnetic vibrations e.g. Take a helix, put in its axis bits of iron wire, heavy glass &c. Send through the helix an intermittent current (best alternately reversed) but the alternations must be very rapid, several billion per sec. Query whether this could be done with a rotating carbon disk & a spring contact a la microphone. Or query whether a steady battery current is not sufficiently intermittent especially if it passes through a vacuum tube. Look at the things in the dark & see if they emit radiation.[19]

He soon realized that the spinning disc would not yield high enough frequencies. The battery and vacuum tube proposal was based on the notion that currents in electrolytes and rarefied gases consist of streams of discrete ions and so are in fact not steady but highly variable, but Lodge soon realized that this kind of intermittency could not be used to generate appreciable waves.

Some of Lodge's other suggestions were more promising. He especially liked the idea of arranging a chain of coils so as to yield high-order induced currents. A series of sudden breaks (a "square wave") fed into the first coil would, he suggested, have its frequency doubled by each succeeding coil; a chain of forty coils with an input of four hundred breaks per second would then yield an output of about 10^{15} cycles per second, enough to generate light in the last coil. Lodge described this plan to FitzGerald in a letter of 26 February 1880 and may have mentioned it at the 1879 British Association meeting as well.[20] But FitzGerald soon pointed out that the coil chain would not work: the doubling would cease after the third or fourth coil as the suddenness of the original breaks was smoothed down into a sine wave.[21]

Lodge had one more idea, however, and it was his best. The discharge of a condenser, he noted, is oscillatory; if its frequency were made high enough, this ought to emit light. Moreover, he wrote in his notebook, "a [Leyden] jar being discharged *does* emit light. This must be the very thing." He squeezed the same suggestion into the margin of his letter to FitzGerald: "I believe the simple discharge of a condenser makes it emit light of this sort: the discharge being oscillatory."[22]

FitzGerald's response is not known; none of his letters to Lodge from this period have survived. But while the oscillations from any condenser

19. Lodge to Larmor, 1 Jan. 1902, OJL-UCL. Note that, following British usage, Lodge's "billions" corresponds to what Americans would call "trillions," i.e., 10^{12}.

20. Ibid.; cf. Lodge to FitzGerald, 26 Feb. 1880, FG-RDS, and Lodge to Larmor, 1 Jan. 1902, OJL-UCL.

21. FitzGerald 1902: 99–101 [1882].

22. Lodge to Larmor, 1 Jan. 1902, OJL-UCL, transcribing part of his 1879–80 notebook; Lodge to FitzGerald, 26 Feb. 1880, FG-RDS.

of appreciable size are in fact far too slow to generate visible light (the light Lodge saw was a secondary effect from the excitation of the molecules of the dielectric), condenser discharges do generate relatively long electromagnetic waves. Indeed, the discharge of condensers was the method later advocated by FitzGerald and used by Hertz for just this purpose. Lodge was coming very close to it in 1879.

FitzGerald and "The Impossibility . . ."

The most important immediate effect of Lodge's work on electromagnetic waves was to stimulate FitzGerald's interest in the problem. In the fall of 1879, apparently after hearing Lodge's British Association paper, FitzGerald asked himself whether it was in fact possible to generate waves in the ether electromagnetically. He almost immediately fell into a serious error that deflected Lodge from his experimental work and confused the whole issue for several years. On the basis of a misapprehension of the implications of Maxwell's theory, FitzGerald concluded that waves could *not* be produced electromagnetically. He stated this in a short paper, "On the Possibility of Originating Wave Disturbances in the Ether by Means of Electric Forces," read before the Royal Dublin Society on 17 November 1879, and repeated his conclusions at the Swansea meeting of the British Association in August 1880 in a paper originally entitled, according to Lodge, "On the Impossibility of Originating Wave Disturbances in the Ether by Electromagnetic Forces." The "Im" was later dropped; "nevertheless," Lodge said, "the answer given was in the negative at that time."[23]

FitzGerald based his mistaken conclusion on two lines of reasoning. First, he argued that the conditions Maxwell had given for the origination of electromagnetic waves could not be met. Maxwell had shown that if we assume that at a given time the vector potential $\mathbf{A}$ and its rate of change $\dot{\mathbf{A}}$ are both zero except in a certain region, then waves will propagate out from that region.[24] But FitzGerald pointed out that if we treat the surrounding space as a nonconductor, such a distribution of the vector potential would violate Gauss's well-known theorem that it is impossible for a potential function to be zero throughout one region and nonzero in another.[25]

23. Lodge to Larmor, 1 Jan. 1902, OJL-UCL; cf. FitzGerald 1902: 90–92 [1879], and FitzGerald, "On the Possibility of Originating Wave-Disturbances in the Ether by Electromagnetic Forces," *BA Report* (1880), p. 497 (title only).

24. Maxwell 1873, art. 785.

25. FitzGerald 1902: 90–92 [1879]; on Gauss's theorem, see Maxwell 1873, art. 144.

This objection broke down when one took into account Maxwell's displacement currents, which were distributed throughout space; Gauss's theorem did not then apply. FitzGerald soon recognized this and brought forward a second and more sweeping argument. In his *Treatise*, Maxwell had repeatedly stated that his theory of electromagnetic action through a medium gave results "mathematically identical" to those derived from the action-at-a-distance theories of Weber and Neumann.[26] But in those theories, FitzGerald pointed out, there is "no account taken of the non-conductor, nor are any variables representing it in any way involved"; there was thus no way, according to those theories, that the energy of an electrical system could be transferred into a nonconducting region (such as the surrounding space), and any closed system of perfect conductors, whatever charges and currents it might include and however it might be moving, must completely conserve its energy.[27] If the identity of results claimed by Maxwell really held, then the same restriction would have to apply to his theory as well, and an electrical system could never give rise to waves that would disperse its energy throughout space. The electrical production of waves like those of light was, FitzGerald concluded, impossible.

FitzGerald had fallen into the trap of taking some of Maxwell's statements too seriously and applying them too widely. Maxwell's theory is in fact *not* always equivalent to the action-at-a-distance theories, and the differences arise in just those cases in which energy is radiated into the field. As Heaviside told FitzGerald nearly ten years later, "You were misled by M's statement of identity of results by German and his methods. But I think he only meant as regards phenomena to be got by *assuming* displacement ignorable."[28] Maxwell's statement really applies only when charges and currents are steady, though he never made this very clear. This was only one of many times that various of the Maxwellians were led astray by obscurities, ambiguities, and outright errors in Maxwell's book. Heaviside said he was "put . . . on the wrong track for years" by Maxwell's mistaken account of how a current rises in a wire; Lodge was put off from possibly discovering the Hall effect by Maxwell's assertion that magnetic forces act only on conductors, not currents; and the list could go on.[29] Extracting a consistent and comprehensive theory from Maxwell's book was among the greatest and most difficult of the Maxwellians' tasks.

FitzGerald persisted in his error for about two and half years, repeat-

26. Maxwell 1873, art. 62; cf. arts. 59, 502, 645.
27. FitzGerald 1902: 90–92 [1879].
28. Heaviside to FitzGerald, 30 Jan. 1889, FG-RDS.
29. Ibid.; Lodge 1931b: 96–97.

ing it in another paper at the Royal Dublin Society in May 1880 and at the British Association that August.[30] Having convinced himself on general grounds that waves could not be produced electrically, he sought in these papers to determine directly what, on Maxwell's theory, the electromagnetic field around a variable current would be. This was a difficult problem, since the displacement currents set up by the original current immediately react on one another, changing their distribution and setting up new displacement currents, and so on ad infinitum. Eventually, however, FitzGerald found a solution that met all the necessary conditions. The resulting field, he wrote, "instead of being such as would initiate wave disturbances like light," would be "more analogous to the nodes and loops in an organ pipe"—that is, a standing wave.[31] The energy in the field would simply oscillate around fixed points, he said; it could never break away and propagate freely across space.

FitzGerald's findings did not lead him to abandon the electromagnetic theory of light; this was, after all, just the time when his long paper on that subject was appearing in the *Philosophical Transactions*. Instead, he took a position similar to the one Chalmers has ascribed to Maxwell. Light, FitzGerald said, was a disturbance in the electromagnetic ether, but it was not generated electromagnetically. He declared in 1879 that "the origination of such disturbances [as light] is not a phenomenon of electric currents such as we have to deal with, but is connected with the relations of matter and ether, and this is probably an atomic interaction, as spectroscopic phenomena also seem to show, and will be explained only when some workable hypothesis as to the nature of this interaction has been sufficiently investigated."[32] As he put it more briefly in his notes, "The interaction between matter and ether that originates light is not the same as that between electric currents and the ether."[33]

For FitzGerald, the origination of light was evidence of a nonelectrical action of matter on ether for which he believed there was other evidence as well. In his theory of the Kerr effect, FitzGerald had (in a muddled way) followed Maxwell in assuming an interaction between molecular vortices and light waves. When he wrote to Stokes to explain and defend this assumption on 17 November 1879—the same day he gave his first paper at the Royal Dublin Society on the impossibility of generating waves electrically—he mentioned that he had "just read a paper at the

30. FitzGerald 1902: 93–98 [1880]; cf. *BA Report* (1880), p. 497.

31. FitzGerald 1902: 93–98 [1880]; the quotation is from a manuscript abstract (in FG-TCD Physics) of the papers FitzGerald presented to the British Association in 1880.

32. FitzGerald 1902: 92 [1879]; Chalmers 1974: 476 quotes this passage to show that the view he ascribes to Maxwell "did not appear as an obviously absurd one at the time."

33. MS abstract (see n. 31), FG-TCD Physics.

R. D. S. that tends to show that the origination of wave disturbances in the ether would require some such interaction of matter and ether as is assumed here so that I have hopes of showing that this is another phenomenon warranting the assumption."[34] It certainly complicated matters to have to regard the origination of light as nonelectrical, but in 1879–80, FitzGerald believed there was good evidence that such was the case.

When Lodge wrote to FitzGerald in February 1880 about his plans for generating "electromagnetic light," FitzGerald apparently tried to throw cold water on the whole idea.[35] Although FitzGerald's reply has not survived, it is clear from Lodge's next letter that FitzGerald argued that light was a bodily undulation of the ether, whereas electromagnetic disturbances were of some different and unknown type—a view Lodge then tried to fit with his own notions of the ether. "It has for some time seemed to me pretty clear," Lodge wrote, "that we have no direct way of acting on or moving the *ether*." Citing his idea that the ether was composed of positive and negative electricity bound together (as in his cogwheel model), Lodge suggested that "when an EMF [electromotive force] is applied the Ether is sheared but not moved bodily either one way or the other. . . . Hence *if* light is bodily undulations of ether then we can't set it going electrically because all Electrical forces urge $+E$ one way and $-E$ the other."[36] This idea, Lodge told FitzGerald, "seems . . . to agree with what you say."

FitzGerald did not endorse the "positive and negative" ether model, but Lodge's letter shows that FitzGerald tried to reconcile the electromagnetic theory of light with his conclusions about the impossibility of generating waves electromagnetically by drawing at least a slight distinction between optical and electromagnetic phenomena. They were intimately related, both being actions of the same ether, but this did not mean they were identical. Maxwell apparently took a similar tack, as we have seen, and it survived as at least a minor current among British physicists for some years to come.[37]

For the most part, Lodge followed FitzGerald's lead in such matters; in these years he was, as Larmor later put it, "in a sense [FitzGerald's]

34. FitzGerald to Stokes, 17 Nov. 1879, GGS-ULC.
35. Lodge to FitzGerald, 26 Feb. 1880, FG-RDS.
36. Ibid. 29 Feb. [1880].
37. See J. H. Poynting to R. T. Glazebrook, 17 Sept. 1893, forwarded to Lodge with a note by Glazebrook, in OJL-UCL. Poynting said he sometimes thought that "we have swallowed the Electromagnetic Theory of Light a little too quickly—that EM waves may be quite different in kind from light waves though in some way using the same qualities of the medium and travelling therefore at the same rate"; Glazebrook agreed.

pupil" on questions of optical and electromagnetic theory.[38] Lodge was well aware of the gaps in his knowledge—he was, he told FitzGerald in 1880, "brutally ignorant of the ordinary undulatory theory at present"—and he was more than willing to draw on his friend's superior training. When he asked for a copy of one of FitzGerald's papers, saying, "I want to get hold of the EM theory of light as it is evidently the true one," he admitted, "I shall not understand your paper . . . at first," adding hopefully, "but I shall grow to it."[39] When FitzGerald declared that ether waves could not be generated electromagnetically, Lodge acquiesced, whether because he fully understood and accepted the argument or because he simply deferred to FitzGerald's authority, it is difficult now to say. Published remarks he made late in 1880 indicate that he at least tentatively accepted FitzGerald's view.[40] In any case, Lodge said later that FitzGerald's mistaken conclusions had "deflected my attention from the subject"; he gave up his efforts to produce ether waves in the laboratory and did not return to the problem until early in 1888, just before news of Hertz's work reached England.[41]

Although FitzGerald's error had a serious effect on Lodge, it apparently attracted little other attention. The *Scientific Transactions of the Royal Dublin Society*, in which FitzGerald's findings first appeared, have never had a wide readership; probably the only scientists outside Ireland to see the papers were those to whom FitzGerald sent offprints, something he never did very systematically. Even Lodge did not see many of these RDS papers until they were collected after FitzGerald's death.[42] For years FitzGerald deliberately followed this course of "burying [his] work in the journals of a provincial society" in an effort to build up the scientific standing of the RDS, until he finally gave up in disgust in 1889 after one of his attempts to reform the society was defeated.[43] His 1880 British Association paper may have been intended to give his ideas wider circulation, but it was in fact not published in the annual *Report* and received only a one-sentence summary in the account of the meeting in *Nature*, which simply stated that by comparing Maxwell's theory to the action-at-a-distance theories, FitzGerald had "deduced the conclusion

38. Larmor 1901b, in FitzGerald 1902: li.
39. Lodge to FitzGerald, 29 Feb. [1880], FG-RDS.
40. Lodge 1881: 303. When this was reprinted, the passage mentioning FitzGerald's apparent proof that electromagnetic waves "cannot be generated by any known electrical means" was deleted; see Lodge 1889b: 319.
41. Lodge to Larmor, 1 Jan. 1902, OJL-UCL.
42. Ibid., 29 Dec. 1901.
43. FitzGerald to R. J. Moss [RDS registrar], 7 Feb. 1889, FG-RDS. The quotation is from a draft; the final letter was published in the *Proceedings of the Royal Dublin Society 125* (1889): 21–23; cf. Mollan 1981: 210–11.

that electric currents and systems cannot originate in the ether such disturbances as those of light."[44] Nor did many hear the paper in person;
the association's Swansea meeting was its most sparsely attended in many
years.

Nonetheless, word of FitzGerald's work reached at least a few British
physicists, and it did not go wholly unchallenged. FitzGerald's friend
John Perry wrote from London in February 1881 to ask for some offprints FitzGerald had promised but not yet sent. Perry and his electrical
engineering partner William Ayrton were, he said, "badly in want of
some information as to what you said at Swansea. I am all at sea on the
matter. I thought I knew Maxwell's theory and then I thought you were
contradicting Maxwell and then Lodge said something agreeing with
Maxwell and you as if it were all clear to him—and here am I wanting to
know what you really said and very much needing the information and
Ayrton in the like condition and not a scrap of information to be had."[45]
Perry thought FitzGerald's work might have some relevance to a "pho-
tophone" he and Ayrton were developing (it did not), but he was also
interested in the general theoretical question. When he got FitzGerald's
papers, he found their arguments unconvincing. Perry thought that "if
Maxwell had seen the difficulty due to the fact that his initial condition is
impossible according to Gauss, he would have altered his theory," perhaps by assuming that potentials were propagated with some large but
finite velocity, a step that would obviate the strictures of Gauss's theorem
without otherwise seriously affecting Maxwell's equations. Perry regarded the argument FitzGerald had founded on the supposed identity
of Maxwell's results and those of the action-at-a-distance theories as similarly inconclusive. The electromagnetic production of light "seems to
me such a probable occurrence," Perry said, that "I should be very sorry
to agree with your deduction from Maxwell's theory."[46]

FitzGerald's response to these remarks has not survived, but his position was clearly becoming unsatisfactory in several ways. Perry had
thought FitzGerald was "contradicting Maxwell," and no doubt others
thought likewise; not everyone was as adept as Lodge at reconciling the
two. As Maxwell's theory was explored more fully in the 1880s—not
least by FitzGerald himself—the "proofs" that it was impossible to generate waves electromagnetically began to look less and less convincing. A
better appreciation of the role of Maxwell's displacement currents and
of the distribution and flow of energy in the field helped rob

44. *Nature 22* (1880): 446.
45. Perry to FitzGerald, 26 Feb. 1881, FG-RDS.
46. Ibid., 6 March.

FitzGerald's arguments of their force. Gauss's theorem did not really apply in the circumstances Maxwell had described, and it became increasingly clear that, despite Maxwell's own claims, his theory did not always give the same results as the action-at-a-distance theories. Moreover, FitzGerald's attempt to show directly from Maxwell's equations that an oscillating current would not radiate energy was seriously deficient, as he soon realized. The "standing wave" he had found in 1880 was a legitimate solution of the relevant differential equation but not the most general one. FitzGerald had in fact shown only that a set of standing waves was a *possible* state of the field around an oscillating current; there might well be others. Thus none of his "proofs" that waves could not be generated electromagnetically were really conclusive. As this became clear in the early 1880s, the time grew ripe for a reassessment of the whole problem.

The Undetected Waves

It was FitzGerald himself who found the flaws in his "proofs." He then proceeded to put the subject on a sound theoretical basis, so that by 1883 he understood quite clearly how electromagnetic waves could be produced and what their characteristics would be. But the waves remained experimentally inaccessible; FitzGerald, along with everyone else, was stymied by the lack of any way to detect them.

Prompted by a passage he had come across in Lord Rayleigh's *Theory of Sound*, FitzGerald began early in 1882 to take a new look at the question of wave production. In treating the origination of sound waves, Rayleigh had dealt with the same differential equation FitzGerald had faced in 1880. But Rayleigh gave a quite different solution, corresponding to a train of progressive waves rather than the standing waves FitzGerald had found.[47] Rayleigh's solution forced FitzGerald to rethink the whole question of electromagnetic wave production, and in "Corrections and Additions," a brief note he presented to the Royal Dublin Society in May 1882, he recanted most of his earlier conclusions.

FitzGerald began by noting that "taking Lord Rayleigh's form of solution would lead to the conclusion that a simply periodic current would originate wave disturbances such as light, and not the stationary waves that my solution leads to."[48] Both were perfectly legitimate solutions, and there was no purely mathematical reason to prefer one over the

47. Rayleigh 1877, arts. 276–78.
48. FitzGerald 1902: 99–101 [1882].

other. But physically it was clear that, for sound waves, Rayleigh's solution applied for the ordinary case of a vibrating body in air, whereas FitzGerald's standing-wave solution would apply only if a reflecting barrier were placed at the proper distance from the source, so that the radiated and reflected waves were resonantly combined. Thus, FitzGerald noted, "If the electromagnetic action is analogous to this, it seems to follow that, in infinite space, where there is nothing to produce the reflected wave, [Rayleigh's] form of solution, in which all the energy is gradually transferred to the medium, is the right one to employ." That is, an oscillating current *would* radiate electromagnetic waves, and Fitz-Gerald's proofs to the contrary must be erroneous.

The proof FitzGerald had based on Gauss's theorem was by this time rather threadbare anyway. Indeed, FitzGerald pointed out that it was refuted even if one took the standing-wave solution he had given in 1880. In that solution, the vector potential was zero at the series of wave nodes, whereas if Gauss's theorem applied, the potential "could never . . . vanish short of infinity." The conclusion seemed inescapable that Gauss's theorem did not apply and, as a corollary, that Maxwell's theory and the action-at-a-distance theories were not really equivalent. This latter point came out even more forcefully when one used Rayleigh's solution, which indicated that an oscillating current would radiate its energy into the surrounding space, directly contradicting the action-at-a-distance theories. It had become clear, FitzGerald noted, that "the assumption that the two theories lead to the same results is only true to the same order of approximation as omitting the mutual induction of the displacement currents in the non-conducting medium." Once these displacement currents were taken into account, one found that Maxwell's theory predicted that *any* oscillating current would produce electromagnetic waves; they became ubiquitous.

FitzGerald realized that this fact had wide implications. For one thing, it eliminated the need for a nonelectromagnetic mechanism to explain the generation of light. As he noted in 1882, "It seems . . . probable that, contrary to what I have stated in my first paper, the interactions between the molecules of matter and the ether are of the same character as the electromagnetic actions with which we are acquainted."[49] Electromagnetic theory could thus be brought to bear on molecular and spectroscopic questions, further extending the unification of optics and electromagnetism that Maxwell had begun.

The realization that oscillating currents could generate waves in the ether also opened up the whole spectrum of electromagnetic waves

49. Ibid., p. 101.

longer than those of light. Once he had cleared up his earlier misapprehensions, FitzGerald recognized that it should be possible to produce waves of virtually any length one wanted—indeed, that they were being produced all the time by ordinary electrical devices. In a July 1882 letter to Henry Rowland, FitzGerald mentioned that he had corrected his earlier error and now found that "it is most likely that all periodic electric and magnetic [oscillations] are accompanied by a loss of energy just like, and in fact the same thing as, the radiation of heat. In fact it extends the dissipation of energy by radiation to the case of periodically working electric and electro-magnetic machines."[50] This conclusion raised the important and difficult question of just how much energy was radiated by an oscillating current. FitzGerald began to attack this question early in 1883, and his notebooks for that year are filled with his struggles to find an answer. The "Poynting flux" had not yet been formulated, and the whole question of the distribution and flow of energy in the electromagnetic field was still largely unexplored. FitzGerald tried many avenues, most of them dead ends, before he found a reasonable solution.

In this work FitzGerald drew heavily on analogies with sound waves and on Rayleigh's *Theory of Sound*. "I must look up Ld. Rayleigh on bells communicating sound &c," he wrote in a notebook entry on the radiation of energy.[51] One of the major puzzles was to determine what part of the energy of an oscillating current was in fact radiated and what part merely sloshed back and forth between the current and the ether. FitzGerald compared it with a problem in the theory of sound: "It is obvious that if a bubble in a big jelly expands or rotates it gives pot[ential] energy to the jelly which when it contracts is restored to it if the changes go on slowly while if they go on fast energy of vibration i.e. kinetic energy will be given to the jelly and it will not in general be restored but will be radiated. Hence the energy of the radiation will depend on how fast the bubble changes i.e. on the period of vibration." Similarly, it was clear that in the electrical case, only a fraction of the energy in the field would be radiated, since "on each reversal a large part of the energy of the ether reverts to the current and it is evidently only the part out of reach of the current during the opposite phase that is lost i.e. the total energy that cannot send back a reverse wave in time to help the reversing current."[52] FitzGerald was here drawing a distinction between the "near field" of a source and its "far field" and showing clearly why this distinction was

50. FitzGerald to Rowland, 13 July 1882, HAR-JHU.
51. FitzGerald notebook 10375, p. 55, FG-TCD Library.
52. Ibid.

important. The key step toward a proper understanding of electromagnetic radiation was to separate the near and far fields and to focus on the energy that escaped from the immediate vicinity of the source. The chief remaining problem was to find how this radiated energy depended on the frequency.

FitzGerald continued to draw on Rayleigh's *Theory of Sound* in working out the equations for energy radiation. For instance, Rayleigh was apparently his source for the idea of "retarded potentials," which later came to play a prominent role in electromagnetic theory.[53] In the same pages in which he had solved the differential equation for a sound source (and so put FitzGerald on the right track about how to generate electromagnetic waves), Rayleigh had used retarded potentials to treat the propagation of waves.[54] In this technique, the conditions at time t at a point a distance r from a source are made to depend on the state of the source at the earlier time $t - r/c$, where c is the speed with which disturbances travel in the medium. In other words, there is a lag between when something happens at the source and when it is felt at a distant point, equal to the time it takes a disturbance to reach that point. FitzGerald applied this treatment to the electromagnetic case, using the retarded vector potential to calculate the energy of the displacement currents set up in the ether by a small oscillating current.[55]

In using retarded potentials, FitzGerald may have been drawing on the work of the Danish physicist L. V. Lorenz, who had introduced them into electromagnetic theory in 1867. Lorenz's paper had been translated (by Lodge's old friend Carey Foster) and published in the *Philosophical Magazine,* so it was readily available.[56] Moreover, FitzGerald later became a strong advocate of Lorenz's methods. Writing to Larmor in 1897 about H. A. Lorentz's electron theory, FitzGerald asked, "Have you seen the other (Copenhagen?) Lorenz's simultaneous-with-Maxwell's-work with these f(t − r/c) functions? It is quite interesting. He entirely escaped the muddle in Maxwell about the forces at one time obeying $\Delta^2 = 0$ and at another $\Delta^2 = a^2 d^2/dt^2$."[57] FitzGerald later said that Lorenz's functions were "essentially the same as I have been always using."[58] Indeed, in the

53. Whittaker 1951: 268–69, 325–26. For a modern discussion of retarded potentials, see Feynman 1964, 2:21.1–21.13.

54. Rayleigh 1877, arts. 276–78.

55. FitzGerald 1902: 122–26 [1883].

56. Lorenz 1867; Whittaker 1951: 268; Pihl 1972; cf. G. C. Foster to Lodge, 23 July 1884 and 19 Sept. 1889, OJL-UCL. Lorenz sought to incorporate an electromagnetic theory of light into his action-at-a-distance electrodynamics, but the physical basis of his theory was self-contradictory.

57. FitzGerald to Larmor, 21 Nov. 1897, JL-RS.

58. FitzGerald to Lodge, 16 Jan. 1901, OJL-UCL. This was one of the last letters FitzGerald wrote. It includes marginal notes by Larmor and is filed among Larmor's letters to Lodge.

1890s, FitzGerald became known as a leading proponent of retarded potentials, and Heaviside (who preferred to call them "progressive" potentials) credited him with having introduced them into electromagnetic theory.[59] In fact the priority clearly belongs to Lorenz, but on the whole it seems likely that in 1883, FitzGerald was unaware of Lorenz's work (he did not cite or mention it until the 1890s) and that he introduced retarded potentials into electromagnetism independently, drawing the basic idea from Rayleigh.

Even armed with the retarded potential method and with Rayleigh's work on sound as a model, FitzGerald faced a hard struggle before he was able to solve the energy radiation problem, and his notebooks for 1883 are filled with queries, question marks, and crossed-out paragraphs.[60] Eventually, however, he found a formula for the energy at a distance r from the source that was separable into two terms: one proportional to $1/r^2$ and one independent of r. The first term is negligible more than a few wavelengths from the source; it represents, FitzGerald said, "the energy of the forced displacement currents produced directly by the variation of the primary current," energy that merely sloshes back and forth between the current and the field.[61] But the second term represents the energy of the additional displacement currents set up by these forced displacement currents; this energy is unable to return to the wire before the primary current reverses, and so is radiated out into space. This flow of energy is real and is the true measure of the energy radiated. FitzGerald's most important result was to find how this term depended on the strength and frequency of the source. He found the radiated power to be

$$P = (\pi a^2 i_o)^2 \, 8\mu\pi^4/3T^4c^3,$$

where a is the radius of the current loop, i_o the amplitude of the current, c the velocity of light, and T the period of oscillation. FitzGerald realized that with c^3, a very large number, in the denominator, the radiated power would be "very small indeed, unless the period T be excessively small."[62] But the $1/T^4$ factor showed that the power increased very rapidly with the frequency, so that at perhaps ten million cycles per second there might be appreciable radiation—and such frequencies could be achieved by discharging condensers. FitzGerald's work, after a long detour, was finally leading him back to Lodge's idea of producing and detecting electromagnetic waves.

59. Heaviside 1893–1912, 3:452 [1910].
60. FitzGerald notebook 10375, p. 76, FG-TCD Library.
61. FitzGerald 1902: 122–26 [1883].
62. Ibid., pp. 128–29 [1883]; also pub. in *Elec. 11* (1883): 518.

FitzGerald presented his findings at the September 1883 Southport meeting of the British Association in a short paper "On the Energy Lost by Radiation from Alternating Currents" and gave a fuller account two months later at the Royal Dublin Society.[63] He also gave another paper at Southport, "On a Method of Producing Electromagnetic Disturbances of Comparatively Short Wavelengths," the published version of which is short enough to reproduce here in full: "This is by utilizing the alternating currents produced when an accumulator is discharged through a small resistance. It would be possible to produce waves of as little as 10 metres wavelength, or even less."[64] From an August 1883 entry in one of his notebooks, it appears that FitzGerald originally planned to title this "On a Method of Producing Sufficiently Short Ether Waves To Be Observable."[65] He had clearly set his sights on producing experimental evidence that electromagnetic waves actually existed.

FitzGerald mentioned the possibility of using condenser discharges to generate waves in his 1882 paper in a reference to Lodge's ideas on "electromagnetic light." He pointed out the flaw in Lodge's coil-chain idea and suggested condenser discharges as a more promising source of the extremely high frequencies required for light—an idea that had, of course, already occurred to Lodge.[66] But FitzGerald's most important step was to look beyond generating light and toward the possibility of producing and detecting much longer waves, for this not only opened up the whole electromagnetic spectrum, which has since proven of such enormous practical value, but also opened the way to a particularly clear and decisive confirmation of Maxwell's theory.

Vague suggestions of waves longer than those of light appeared in FitzGerald's 1882 paper, but he first made the idea explicit in notebook entries dating from early 1883. In a section headed "Period of free vib[ration] of a current," dated 9 April 1883, he calculated the oscillation period for a simple circuit containing capacitance, inductance, and resistance and examined how these could be adjusted to give the most rapid oscillations and so the most energetic radiation. At first he thought he could produce high enough frequencies "that there seems just a possibility of getting such vibrations visible even," but he had been misled by a numerical error.[67] A recalculation showed that wavelengths of a few centimeters were the practical minimum, and of a few meters, as he

63. Ibid.; FitzGerald 1902: 122–26 [1883].
64. FitzGerald 1902: 129 [1883]; also pub. in *Elec. 11* (1883): 519. By "accumulator," FitzGerald meant a condenser.
65. FitzGerald notebook 10375, p. 74, FG-TCD Library, in a list of the titles of other papers FitzGerald presented at the 1883 BA meeting.
66. FitzGerald 1902: 100 [1882].
67. FitzGerald notebook 10375, pp. 56–58, FG-TCD Library.

mentioned at the 1883 British Association meeting, a more reasonable experimental goal.

This limitation pointed up, however, the basic problem with using long waves: how were they to be detected? For light, we can use our eyes, but we have no organs sensitive to longer waves; nor did any of the scientific instruments available in the early 1880s seem capable of detecting them. FitzGerald saw early on that the key to detecting long waves would be to use the principle of interference. "By the method of interference," he observed, "you can find out if there are waves at all and by testing near a single coil you can find whether there are progressive or stationary waves."[68] One would simply reflect the waves from a wall and adjust the distance until a resonance was reached and then, by searching out the nodes and loops, proceed to measure the wavelength directly. The method was well known from experiments on sound and was in fact exactly the one Hertz was to use in 1888.

But FitzGerald still had to find a way to detect the waves. He could bring them to a stop, so to speak, so that they formed a standing interference pattern instead of whizzing by at the speed of light, but he still had no instrument able to respond to the high frequencies—millions of cycles per second—at which the standing waves would oscillate. He thought in 1883 that "by getting a sensitive instrument like an induction balance on a large scale we might produce interference of the waves," but he soon realized that induction balances would not work for such high frequencies, nor could he think of any detector that would.[69] More than a year later, the problem was still bothering him. When he wrote to J. J. Thomson in December 1884 to congratulate him on being named Cavendish Professor at Cambridge (a post FitzGerald had himself sought in a rather half-hearted way), he added, "I hope you will succeed in getting your experimental test of Maxwell's theory tried. The great difficulty is something to *feel* these rapidly alternating currents with. Would Langley's bolometer do? I was working at a receiver whose period of oscillation should be the same as that of the current and which would consequently 'resound' to the vibration and integrate the energy of a large number of vibrations."[70] Langley's bolometer was a sensitive thermocouple for measuring radiant energy; the idea was to detect the heat generated by the rapidly alternating currents induced by the incident waves, but calculations with FitzGerald's energy formula suggested that with practicable sources the effect would be undetectably small. The idea of a receiver that would "resound" was far more promising; Hertz's de-

68. Ibid., p. 58.
69. Ibid., p. 57.
70. FitzGerald to J. J. Thomson, 23 Dec. 1884, in Rayleigh 1943: 21–22.

tector depended on resonance, as have virtually all electromagnetic wave detectors developed since. The resonant detector FitzGerald said he had been working on was probably a very sensitive galvanometer or electrodynamometer with its natural oscillation period adjusted to match that of the waves; a short paper he published in June 1884 was probably an outgrowth of this work.[71]

But in fact neither FitzGerald nor Lodge nor any other scientist actually built a wave detector before 1888.[72] Lodge said later that in light of all he and FitzGerald knew about electromagnetic waves by 1883–84— how to produce them by discharging condensers and what their wavelength and energy would be— "the wonder is that we neither of us followed this up with vigour and determination."[73] In fact, however, FitzGerald rarely followed up such things "with vigour and determination"; he was mainly an idea man, known for tossing off suggestions in hopes that others would carry them out. Rather than performing experiments himself, he generally left them to his students and assistants, of whom he later had outstanding ones in John Joly, Frederick Trouton, and Thomas Preston.[74] But in the early 1880s he had not yet acquired this circle of followers in Dublin, and his ideas for experiments were likely to be carried out only if they were taken up by one of his friends in England—notably Lodge. Lodge, however, was extremely busy with other work in these years, especially after he moved to University College Liverpool in 1881. As one of the first professors at the new institution, he carried an enormous teaching load in the early years and sometimes lectured for as much as five hours a day; "Even standing up all that time was tiring," he later said. In addition, "It was no joke having to start a laboratory from the beginnings," and it was only with great effort that he found any time for his own research.[75] Lodge worked in two main areas in the early and mid-1880s: electrolysis, particularly the speed of ionic motion, and the electrostatic deposition of dust. Both were successful (dust deposition at smelters later became a big business, run by Lodge's son Lionel) and helped establish Lodge as a promising young physicist, but he was left with little time in which to ponder the ether or search for electromagnetic waves.[76]

71. FitzGerald 1902: 134–35 [1884].
72. Several technologists, including Thomas Edison and D. E. Hughes, detected what are now thought to have been electromagnetic waves in the 1870s and early 1880s but did not recognize their wave nature or their theoretical significance; see Susskind 1964.
73. Lodge 1931a: 90.
74. On Joly and Trouton, see their *DSB* entries; on Preston, see Weaire and O'Connor 1987.
75. Lodge 1931b: 159, 153.
76. Lodge 1931a: 59–67, 79–80.

Even if FitzGerald and Lodge had pursued the waves vigorously, there is no assurance they would have found them. For one thing, FitzGerald's proposed wire-loop oscillator was, as Lodge later noted, "a shocking bad radiator"; because it emitted energy from the weak magnetic field around the wire rather than from the strong electric field between the condenser plates, it would have required very large internal currents and voltages to produce any appreciable radiation.[77] Moreover, none of the detectors that either FitzGerald or Lodge proposed in the early 1880s would have worked, nor is there any evidence that they were on the track of either the method Hertz used in 1888 or any of the others developed later. Hertz himself said later that, admirable as FitzGerald's theoretical work was, he did not think that "theory alone" could have led to the actual discovery of electromagnetic waves; "For their appearance upon the scene of our experiments depends not only upon their theoretical possibility, but also upon a special and surprising property of the electric spark which could not be foreseen by any theory."[78] FitzGerald and Lodge never suspected that electromagnetic waves would be powerful enough to cause visible sparks to jump across a gap in a wire loop in the way Hertz observed. They thus missed a major opportunity to confirm their own ideas about the waves, and Maxwell's theory as a whole.

FitzGerald and Lodge did not succeed in their early efforts to detect electromagnetic waves experimentally, but by the mid-1880s they had made substantial progress in exploring this crucial aspect of Maxwell's theory. Lodge had launched the idea of producing "electromagnetic light" and had spurred his friend FitzGerald to look into the question; FitzGerald, after surmounting some initial misapprehensions, had gone on to extend Lodge's idea to electromagnetic waves of any length and to work out their main properties mathematically. The theory of electromagnetic waves, which was barely sketched in Maxwell's *Treatise* and there substantially limited to light waves, was by 1883 a soundly based and rapidly advancing part of the research front. Through the rest of the 1880s, work on the propagation of electromagnetic waves played a leading part in the consolidation of the Maxwellian synthesis and, not incidentally, of the Maxwellian group as well.

77. Lodge to Larmor, 29 Dec. 1901, OJL-UCL; this is the second of two letters of this date, both later returned with Larmor's marginal comments. Although FitzGerald could presumably have calculated the radiative advantage of using more widely separated plates, neither he nor Lodge apparently appreciated this point until after Hertz had demonstrated it experimentally in 1888. This point is discussed more fully by Jed Buchwald in a forthcoming book on the work of Hertz.
78. Hertz 1893: 3.

CHAPTER THREE

Heaviside the Telegrapher

FitzGerald and Lodge were both physics professors, and they tended to approach Maxwell's theory and the question of electromagnetic propagation as matters of essentially academic interest. Oliver Heaviside, who was to join them as one of the central figures in the Maxwellian group, came to the subject from a very different background and with somewhat different concerns. He left school at age sixteen and never held an academic position; for most of his life he held no job at all. He has been called "the last amateur of science," and in the sense that he worked independently and largely without pay, Heaviside was indeed an amateur.[1] But his motivations for studying electromagnetism were in fact very practical: he began his electrical career as a telegrapher and turned to Maxwell's *Treatise* and the theory of electromagnetic waves in the 1870s in hopes of finding a way to improve signaling along submarine cables. From its beginnings in the 1850s, submarine telegraphy had played an important part in the evolution of British thinking about electromagnetism; in Heaviside's work, this interaction between technology and theory reached its highest point.

Heaviside eventually emerged as the most penetrating and influential student of Maxwell's theory; it was in fact he who cast it into the form now universally known as "Maxwell's equations." He was guided throughout by questions drawn from telegraphy, particularly those involving electromagnetic propagation; these questions effectively determined the course of his research and are the key to understanding the

1. Cardwell 1972: 124n.

evolution of his work. As W. E. Sumpner aptly observed, Heaviside "was a mathematician at one moment, and a physicist at another, but first and last, and all the time, he was a telegraphist."[2]

Oliver Heaviside

Heaviside was an unusual personality; his best friend once described him as "a first-rate oddity," though, he felt compelled to add, "never a mental invalid."[3] A self-centered man with a caustic wit and few social skills, he lived much of his life as a virtual hermit. In part because of the circumstances of his early life, and in part because of personal peculiarities that were probably inseparable from his genius, Heaviside was always an outsider in the British scientific community and in Victorian society as a whole.

Oliver was born 18 May 1850 in London, the fourth son of Thomas and Rachel West Heaviside. The elder Heaviside has been described as "a zealous radical and free-thinker" and "a man of despotic and unsympathetic temper."[4] A skilled wood engraver, he had come to London from Stockton-on-Tees in the 1840s in search of steady work, but his trade was being eroded by the introduction of photographic etching techniques, and the family was chronically short of money. To help make ends meet, Heaviside's mother opened a small dame's school in the family's rented rooms in a decaying part of Camden Town in North London.

The harshness of Oliver's early years was compounded by illness, including a bout with scarlet fever that seriously damaged his hearing. As he later told Hertz, "I was very deaf from an early age to manhood, and that has influenced my whole life."[5] Although he recovered most of his hearing by his teens, this early deafness left a permanent mark on his personality. Heaviside remained a loner throughout his life and tended to be suspicious of others, as if he suspected them of talking behind his back.

A letter Heaviside wrote to FitzGerald in 1897 gives a vivid picture of his life in Camden Town:

> I was born and lived 13 years in a very mean street in London, with the beer shop and bakers and grocers and coffee shop right opposite, and the

2. Sumpner 1932: 837.
3. Searle 1950: 96.
4. Whittaker 1929: 200. The fullest account of Heaviside's early life is in Nahin 1988: 13–25; see also Appleyard 1930: 212–20 and Lee 1950: 10–12.
5. Heaviside to Hertz, 13 Sept. 1889, HH-DM.

Oliver Heaviside, early 1890s. Courtesy IEE Archives

"I believe I do right to record the conviction that he was never, at any time, a mental invalid. Of course, he was a first rate oddity—he was Oliver."

 —G. F. C. Searle, recalling his friend Oliver Heaviside, 1950

ragged school and the sweeps just round the corner. Though born and bred up in it, I never took to it, and was very miserable there, all the more so because I was so exceedingly deaf that I couldn't go and make friends with the other boys and play about and enjoy myself. And I got to hate the ways of tradespeople, having to fetch the things, and seeing all their tricks. The sight of the boozing in the pub made me a teetotaller for life. And it was equally bad indoors. [Father was] a naturally passionate man, soured by disappointment, always whacking us, so it seemed. Mother similarly soured, by the worry of keeping a school.[6]

Heaviside's childhood was almost literally Dickensian; when Dickens was a boy and working miserably in a London blacking factory, he had lived in lodgings not far from where the Heavisides later settled. "I used to live with Dickens, with his characters, that is to say," Heaviside told FitzGerald, remarking on another occasion, "I know exactly how he got his most intimate knowledge of the lower middle class, as shown in his early papers."[7] He was a great reader of Dickens's novels, and the names of "Sam Weller," "Mrs. Gamp," and other Dickens characters are sprinkled through his books and letters. These references were humorous, but the wit was biting, and behind it lay a bitterness rooted in Heaviside's experiences as a poor, deaf boy growing up in Camden Town.

When Oliver was thirteen, a small legacy enabled his family to move to a better part of Camden Town. "It was like heaven in comparison," he said later, "and I began to live at once."[8] He was sent to the local Camden House grammar school and did especially well in the natural sciences, winning an examination medal in 1865.[9] His parents could not afford to keep him in school much longer, however, and his formal education came to an end the next year.

Heaviside's early schooling in science was not notably deficient; while not as good as FitzGerald's, it was better than that received by Lodge or most other well-to-do English boys of the day, who were taught much Latin and little or no science. But the most important part of Heaviside's education, as of Lodge's, took place out of school. The scope of his reading, and of his mathematical talent, is suggested by one of the characteristic autobiographical parables he wrote in 1902:

More than a third part of century ago, in the library of an ancient town, a youth might have been seen tasting the sweets of knowledge to see how he liked them. . . . In his father's house were not many books, so it was like a

6. Heaviside to FitzGerald, 3 June 1897, FG-RDS.
7. Ibid., and Heaviside to FitzGerald, 8 May 1899, FG-RDS.
8. Heaviside to FitzGerald, 3 June 1897, FG-RDS.
9. Nahin 1988: 17; Lee 1950: 11; Whittaker 1929: 200.

journey into strange lands to go book-tasting. Some books were poison; theology and metaphysics in particular; they were shut up with a bang. But scientific works were better; there was some sense in seeking the laws of God by observation and experiment, and by reasoning founded thereon. Some very big books bearing stupendous names, such as Newton, Laplace, and so on, attracted his attention. On examination, he concluded that he could understand them if he tried.

But other big books, labeled "Quaternions," he found incomprehensible, and "after the deepest research, the youth gave it up, and returned the books. He then died, and was never seen again."[10] How closely Heaviside's experiences paralleled those of "the deceased youth" is hard to say; certainly his introduction to quaternions, from which he later derived his vector system, took a different form. But it is clear that by the time he was sixteen or seventeen, Heaviside had, largely by private study, laid the foundations of an extraordinary scientific education.

With a university education beyond his means, Heaviside faced the necessity of finding a job. After about a year spent reading and otherwise preparing himself, he headed north to Newcastle in 1867 to begin work as a telegrapher. It was a step that set the course of almost all his later work. That Heaviside chose telegraphy was the result of a fortunate family connection: one of his uncles by marriage was Sir Charles Wheatstone (1802–75), inventor of the telegraph. Originally a musical instrument maker, Wheatstone became well known in the 1820s for his scientific studies of sound propagation; after 1830 his attention shifted to electricity, and in 1834 he was named professor of physics at King's College London. His development with W. F. Cooke of a practical electric telegraph won him a substantial fortune and later a knighthood. In 1847 he married Emma West, an older sister of Heaviside's mother, and settled first at Hammersmith and later near Regent's Park in London.[11]

As Wheatstone's "poor relations," the Heavisides looked to him for help in finding careers for their sons, and he obliged: one of Oliver's older brothers, Charles, became an expert on the concertina (another of Wheatstone's inventions) and went into the music business; another, Arthur, became Wheatstone's assistant and later ran one of his local telegraph companies in Newcastle. After the British inland telegraphs were nationalized in 1870, Arthur became a leading engineer in the Post Office telegraph system.[12]

---

10. Heaviside 1893–1912, 3:135–36 [1902].
11. Bowers 1975: 155–56. Bowers notes that the Wheatstones' first son, Charles Pablo, was born just three months after their marriage.
12. "A. W. Heaviside" [obituary], *JIEE 61* (1923): 1154–55.

Oliver was originally sent north to serve as Arthur's assistant, but in the fall of 1868 he was hired to work on the newly laid Anglo-Danish cable running from Newcastle to Denmark, first as an operator and later as an "electrician," as experts in electrical theory or practice were then called. He spent the next several years in testing rooms and on cable ships, learning firsthand the ins and outs of submarine telegraphy, then the most advanced and scientifically interesting branch of electrical technology.

The Cable Empire

Submarine telegraphy was still a relatively young industry when Heaviside entered it in the 1860s, but it had already begun to change the world. The first successful cable had been laid from Dover to Calais in 1851, and it was soon followed by others, often laid with more enthusiasm than expertise, across the North Sea, the Irish Sea, and the Mediterranean. Messages that had formerly taken days or weeks to travel by ship now reached their destinations in minutes, and international trade, the dissemination of news, and the conduct of foreign affairs were changed forever. There was excited talk of tying the whole world together with wires, and plans were made to lay cables to India and America. These ambitions suffered a setback in 1858 when the first Atlantic cable failed after just a month of fitful service, but the more carefully laid Atlantic cable of 1866 proved a great success and was soon followed by cables to India, Australia, and Hong Kong. By 1885 nearly one hundred thousand miles of cable had been laid, and almost the entire world had been brought within reach of the telegraph.[13]

The great majority of these cables were laid and operated by British firms. Indeed, after about 1870 most of them were under the control of one man, Sir John Pender, a Manchester textile merchant and Liberal politician who had been an early backer of the Atlantic cable projects. As chairman of both the Telegraph Construction and Maintenance Company and the Eastern Telegraph Company and its affiliates, Pender came to dominate both the manufacture and operation of the world's major cables. About 70 percent of the total cable mileage in place in 1900 had been manufactured and laid by TC&M, and about 75 percent of the international cable routes had been financed by British capital.[14]

13. Coates and Finn 1979: 165; see also Bright 1898.
14. Coates and Finn 1979: 170–71. For a fairly full though sometimes unreliable account of Pender and his enterprises, see Barty-King 1979: 3–109.

There is no great mystery behind Britain's dominance of the world cable network: Britannia ruled the waves in the nineteenth century, and it was only natural that she should seek to extend that rule beneath them as well. As a maritime trading nation and as easily the leading commercial, industrial, and imperial power of the day, Britain had both the greatest need for cables and the greatest capacity for laying them. The web of undersea cables was often called the "nervous system" of the British Empire: information flowed in along the cables and commands flowed out, binding the empire and the entire world more closely together. Cables had obvious military and diplomatic uses, but perhaps their most important contribution was to provide up-to-the-minute news of foreign markets, thus facilitating the smooth operation and growth of the world trading system that sustained Britain's wealth and power. In return, Britain's imperial and commercial power reinforced its dominance of the cable industry in some quite direct ways. British firms were able, for instance, to establish a virtual monopoly on the trade in gutta-percha, a natural plastic derived from the sap of certain East Indian trees which was the favored material for insulating cables. Foreign companies were unable to obtain the hundreds of tons of gutta-percha needed for long cables and so were effectively shut out of competition.[15]

The Anglo-Danish cable that Heaviside joined in 1868 was an exception to the rule of British control: a key link in the "Great Northern" chain that ran across the Baltic and Russia to the Far East, it was owned by a Danish company.[16] Nonetheless, the 350-mile cable across the North Sea was manufactured, laid, and tested by British contractors. By the 1860s, Britain had built up a corps of cable engineers whose knowledge and skills were unrivaled, and British technical expertise came to dominate submarine telegraphy even more thoroughly than did British capital. It was thus British electricians who, far more than those of other countries, were exposed to the phenomena peculiar to this new technology. Simply to make their cables work, they had to place themselves at the forefront of electrical science, and in the 1850s and 1860s the pace and direction of electrical research in Britain was set largely by the needs and resources of submarine telegraphy.

Beginning when he was just eighteen, Heaviside was exposed to this new and evolving technology at its heart: in a cable company testing

15. On the role of cables in the British Empire, see Headrick 1988: 97–144; on gutta-percha, see Bright 1898: 248–331.
16. Bright 1898: 112–16. Though not British, the Great Northern Co. had close ties to Pender's group, including a profit-pooling arrangement; see Ahvenainen 1981: 20–24, 51.

room, where the electrical characteristics of the cable were carefully measured and a close watch kept for the appearance of faults in its insulation. Submarine telegraphy had not yet settled fully into a routine in Heaviside's day, and testing rooms still possessed, as he later said, some of "the spirit of scientific research"; the electricians were not afraid to experiment with new ways of working their cables and to explore the new phenomena they sometimes encountered.[17] Cable testing rooms were among the most sophisticated and best-equipped electrical laboratories in the world in the 1860s, and many instruments and measuring techniques that only later found their way into university physics laboratories had their origins in the work of cable electricians.

One of the most important areas in which cable telegraphy led the way was in the development of standard electrical units. Accurate standards of resistance, current, and potential were vital to the efficient operation of large cable systems, and it was in response to calls from William Thomson and from two prominent cable engineers, Latimer Clark and Sir Charles Bright, that the British Association established its Committee on Electrical Standards in 1861. Over the next decade, this committee brought leading British physicists and engineers together in a common effort to rationalize methods of electrical measurement. Maxwell joined in 1862 and, along with Balfour Stewart and the cable engineer Fleeming Jenkin, undertook the most important of the committee's experimental work, the determination of the standard ohm.[18]

The work of the standards committee had an enormous impact on electrical technology. Neither the extension of the worldwide telegraph system nor the development of the new electric power industry could have been carried through as effectively without the system of ohms, volts, and amperes the committee provided. Scientific research also benefited as the spread of "B. A." units and measuring techniques brought a new clarity and precision to electrical experimentation. In addition, Maxwell's determination of the standard ohm added greatly to the evidence that the so-called ratio of units was equal to the speed of light and so lent important support to his new electromagnetic theory of light.[19]

Maxwell's experience on the standards committee brought home to

17. Heaviside 1893–1912, 1:428 [1893].
18. The committee reports appeared in the annual *BA Reports* between 1862 and 1870 and were reprinted in Jenkin 1873. On the work of the committee, see Smith and Wise 1989: 687–98; on its origins, see Jenkin to William Thomson, J36 and J37, WT-ULC, two undated letters written during the 1861 British Association meeting. A broken leg kept Thomson from attending the meeting, so Jenkin acted on his behalf in organizing the formation of the committee and in seeking the cooperation of Bright and Clark, who had independently issued their own call for a system of standardized units.
19. Everitt 1975: 99–101.

him the importance of the interaction between science and technology, and in the preface to his *Treatise* he made a point of calling attention to the way in which

> the important applications of electromagnetism to telegraphy have . . . reacted on pure science by giving a commercial value to accurate electrical measurements, and by affording to electricians the use of apparatus on a scale which greatly transcends that of any ordinary laboratory. The consequences of this demand for electrical knowledge, and of these experimental opportunities for acquiring it, have been already very great, both in stimulating the energies of advanced electricians, and in diffusing among practical men a degree of accurate knowledge which is likely to conduce to the general scientific progress of the whole engineering profession.[20]

The "demand for electrical knowledge" that cable telegraphy generated made it a driving force in British electrical research in the third quarter of the nineteenth century. Heaviside, in the testing room at Newcastle, was right in the middle of this interplay of technology and science.

At Newcastle

In his later years Heaviside often poked fun at the "stupid old Toryism" of Cambridge mathematicians, but his scorn was mixed with envy; for he knew firsthand how hard it was to make one's own way in the scientific world. "I regret exceedingly not to have had a Cambridge education myself," he said in 1894, "instead of wasting several years of my life in mere drudgery, or little more."[21] Nonetheless, he was able to learn far more about the actual behavior of currents, magnets, and condensers while working at Newcastle than he could have in those days at Cambridge, which had no physics laboratory until 1874 and no undergraduate laboratory classes until much later.[22] Heaviside's stint as a cable electrician was perhaps not as pleasant as three years at a Cambridge college would have been, but it gave him a better grounding in the real phenomena of electromagnetism than he could then have acquired almost anywhere else.

Nor was all of Heaviside's time with the cable company spent in "mere drudgery," though he was worked hard. A journal he kept for a few

20. Maxwell 1873: vii–viii.
21. Heaviside 1893–1912, 3:515 [1901], 2:10 [1894].
22. Physics teaching laboratories had begun to appear at other British universities from the mid-1860s; see Gooday 1989. Significantly, they emphasized precision electrical measurement of the kind pioneered by cable engineers.

years after 1870 shows that he was able to squeeze out time for independent reading and research and that he was continuing his efforts to teach himself physics. Along with accounts of work done on the cable, the journal includes notes of his reading in the latest scientific books and journals, speculations on lightning and the aurora, and accounts of various experiments he had performed.[23]

Although Heaviside was surrounded by other electricians at Newcastle, he remained intellectually isolated. There seems to have been no one, not even his brother, with whom he felt free to discuss the views he was coming to hold about science and the world in general. This comes out clearly in an 1871 journal entry: "How true is Carlyle's observation that when a man, entertaining, as he believes, *solitary* heterodox opinions meets another holding the same opinions, how strongly is his belief strengthened! He is no longer alone! I have today been reading some of Tyndall's Essays on Materialism, Miracles, etc., and what was to me before only a strong opinion, is now Conviction. But not unreasoning conviction."[24] Heaviside retained these heterodox opinions throughout his life; he was essentially a scientific materialist. The workings of the universe are entirely determinate, he said, "even in the minutest particulars in the history of nations or of animalculae, or in the development of a human soul."[25] He regarded the advancement of scientific understanding of the physical world as the only way to banish superstition and "metaphysics," and he was imbued with a sort of missionary zeal to increase and spread such understanding, particularly about his own subject of electricity. In the early 1870s he was, he said later, "a mighty enthusiast, filled with a strong sense of my Duty to impart my knowledge to others and help them," and it was mainly from "philanthropic motives" that he decided to publish his work.[26]

Heaviside's first paper appeared in July 1872 in the *English Mechanic*, a popular weekly journal, over the signature "O." It described a simple but effective method for comparing electromotive forces which he had devised in 1870 while working on the Anglo-Danish cable.[27] In February 1873 he made his debut in the *Philosophical Magazine*, the leading physics journal of the day, with a paper "On the Best Arrangement of Wheatstone's Bridge for Measuring a Given Resistance with a Given Galvanometer and Battery."[28] It was in this paper that he first showed the

23. Heaviside notebook 1a, OH-IEE.
24. Ibid., entry for 4 May 1871.
25. Heaviside 1892, 2:77 [1886].
26. Heaviside to J. S. Highfield [draft], 14 Mar. 1922, OH-IEE; Heaviside 1892, 1:vii.
27. Heaviside 1892, 1:1 [1872]; cf. Heaviside notebook 3a, entry 1, OH-IEE. Notebook 3a is Heaviside's own annotated bibliography of his publications.
28. Heaviside 1892, 1:3–8 [1873].

mathematical tenacity that was to become a hallmark of his later work; for though the Wheatstone bridge was a simple device and widely used by both telegraphers and physicists, the algebraic obstacles to finding its optimal arrangement were formidable. Heaviside's straightforward solution of a problem that had already defeated several better-known scientists brought him to the attention of leaders in the field, including William Thomson and Maxwell. Later Heaviside said, "A very short time after [this paper] appeared, [I] saw Sir W. Thomson at Newcastle, who *mentioned* it, so I gave him a copy. . . . Cuff told me Sir W. said he had tried to work it out, but found the algebra too heavy. S. E. Phillips also congratulated me upon it, as he had tried at it. So paper was a good beginning. Sent Maxwell a copy, and he noted it in his 2nd Edn."[29] Heaviside had a respect "almost amounting to veneration" for Thomson, "on account," he later said, "of his invaluable labours in science, inexhaustible fertility, and immense go," and this meeting with him must have been a high point in the young telegrapher's life.[30] The two men corresponded intermittently for many years thereafter, and on several occasions Thomson gave important help in publicizing Heaviside's work.[31] Heaviside's direct contact with Maxwell seems to have been limited to sending him a copy of his paper, but a reference in one of Maxwell's humorous poems suggests he was aware of at least some of the other work Heaviside published in the 1870s.[32]

Like all telegraph electricians, Heaviside was eager to find ways to increase the amount of traffic his wires could handle. An especially attractive way to double the carrying capacity would be to send signals simultaneously in both directions along a single wire, but despite strenuous efforts from the 1850s on, it was not until 1872 that the American inventor J. B. Stearns showed how such "duplex" signaling could be made fully practical. Aided by his brother Arthur, Heaviside began to experiment with duplex signaling at Newcastle early in 1873, and in June he published a paper "On Duplex Telegraphy," in which he surveyed the methods then in use and proposed two new ones.[33]

It was in "Duplex Telegraphy" that Heaviside first displayed the sarcasm and contempt for authority that were to land him in controversies throughout his career. Among those at whom he poked fun were the

29. Heaviside notebook 3a, entry 3, OH-IEE. Cuff and Phillips were telegraph engineers. Heaviside's result is cited in Maxwell 1881, art. 350.
30. Heaviside 1892, 1:419 [1885].
31. See Gossick 1976. Gossick was apparently unaware that Heaviside and Thomson actually met in 1873.
32. In an 1877 poem, Maxwell mentioned "two Olivers," identified by his biographers as Heaviside and Lodge; see Campbell and Garnett 1882: 374n.
33. Heaviside 1892, 1:18–24 [1873].

"experts" who had declared that signals sent simultaneously in opposite directions would destroy each other "after the manner of the Kilkenny cats". The inventors of the duplex system had "effectually silenced this powerful argument," Heaviside said, "by going and doing it." R. S. Culley, "the very practical author of *Practical Telegraphy*" and engineer-in-chief of the Post Office telegraph system, was subjected to particular ridicule for his persistence in dismissing duplex signaling as impractical.[34] Heaviside's whole paper had an air of confidence, even brashness, that was perhaps a bit jarring, especially coming from a little-known telegrapher who had just turned twenty-three. Others might find it difficult, for instance, to see how a complicated Wheatstone automatic telegraph could be operated in a duplex circuit, "but I have done it in a very simple manner," he said airily, "which it is unnecessary to describe." It is not surprising that Heaviside's writings rubbed some people the wrong way.

One who took especially strong exception to Heaviside's paper was William Henry Preece, then head of the Southern Division of the Post Office Telegraph Department. In 1873 he wrote to tell Culley, his immediate superior, that "Oliver Heaviside has written a most pretentious and impudent paper in the *Philosophical Magazine* for June. He claims to have done everything, even Wheatstone automatic duplex! He must be met somehow." Culley replied, "O. Heaviside shows what is to be done by cheek. . . . We will try to pot Oliver, somehow."[35]

This was only the first of Heaviside's nearly lifelong string of disputes with the Post Office engineers, particularly Preece, who later became chief engineer himself. In the 1880s, Heaviside became convinced (not without reason) that Preece was trying to silence him and suppress his work, especially those parts that contradicted Preece's own pronouncements.[36] Heaviside was soured by the experience, and his tendency to be suspicious of others began, in his attitude toward Preece, to shade into paranoia. Animosity toward the man he called "the eminent scienticulist" became one of the strongest threads running through Heaviside's career.

Late in 1873, Heaviside ran into trouble when he tried to join the new Society of Telegraph Engineers. The society, the progenitor of the present-day Institution of Electrical Engineers, had been founded in London in 1871 and came to Heaviside's attention the next year, when his brother Arthur joined. "A. W. H. said I should join the new Society,"

34. Ibid., p. 18; cf. Culley 1871: 223.
35. Quoted in Baker 1976: 109–10.
36. See below, Chap. 6, and Nahin 1988: 139–87.

Heaviside wrote later. "But there were snobs in those days. . . . On enquiry he was told they didn't want telegraph clerks!"[37] This was enough to get Heaviside "riled," he said; he had a high opinion of his own worth and was always sensitive to suspected slights. He hated to be lumped with mere clerks. Whether this was part of an effort by Culley and Preece to "pot Oliver, somehow" is not clear, but Heaviside was convinced they were behind it. He decided to go over their heads: drawing on the slender store of scientific capital he had built up with his Wheatstone bridge paper, he turned to Sir William Thomson, whom he had met just a short time before, "and asked him to propose me. He was a real gentleman, and agreed at once."[38] Thomson also enlisted the aid of Sir William Siemens, the first president of the society, and with such backing, it is not surprising that Heaviside was admitted, "in spite of the P. O. snobs," in January 1874. He was even elected to the society's council in 1876, but while he published several papers in the society's *Journal*, he rarely if ever attended its meetings. After a few years he stopped paying his dues, and in 1881 he was struck from the list of members.[39]

By then Heaviside was no longer employed as a telegrapher. He had quit his job in Newcastle in May 1874 and returned to his parents' home in London, reportedly because he thought his working conditions left him too little time for research; the move was apparently also prompted by poor health.[40] Oliver's health had been a matter for concern since his childhood bout with scarlet fever, and in 1874 he apparently suffered an attack of the "hot and cold disease" that was to plague him throughout his life. He described its usual course to FitzGerald in 1897: "I took a dreadful chill . . . and it flew to the stomach and bowels as usual, and then flew to the brain, which was blown up over and over again. It is wonderful what the brains will stand; break and mend again. I have often wondered that I am not in a madhouse, incurably imbecile, brains all smashed and mixed up."[41] The illness was unpleasant enough, but Heaviside's main concern, as this letter suggests, was how it might affect his mind. In another letter, he took FitzGerald to task for referring in print to Heaviside's "bodily infirmities"; they were not something he wanted publicized. "My principal infirmity," he said, "is that I am a chronic dyspeptic, and it occasionally gets awful, and brings on nervous disturbances, which may culminate in epilepsy some day, or may not, according as I live it down, or not." Heaviside's mother was epileptic, and

37. Heaviside to Highfield [draft], 14 Mar. 1922, OH-IEE; cf. Heaviside notebook 1a, entry for 30 May 1872, OH-IEE.
38. Heaviside to Highfield [draft], 14 Mar. 1922, OH-IEE.
39. Appleyard 1939: 93–94.
40. Nahin 1988: 22; Baker 1976: 208.
41. Heaviside to FitzGerald, 3 June 1897, FG-RDS.

fear that he too might develop it was probably one of the main reasons, along with his early deafness, that he led such a retired and solitary life.[42]

After recovering from his 1874 illness, Heaviside stayed on with his parents in Camden Town and devoted all his time to private study and research. In the early 1880s, Preece twice offered to recommend him for jobs on the Post Office research staff or with Western Union in America, but Heaviside was suspicious of Preece's motives and turned down the offers.[43] He preferred to pursue his electrical studies alone, without distractions or interference; if this meant living in comparative poverty, that was a price he was willing to pay. "I was born a Natural Philosopher," he said toward the end of his life, "not an active Engineer, nor yet a 'practical man' in the commercial sense."[44] He was temperamentally unsuited to the jobs he was offered, perhaps to any ordinary job. Arthur Heaviside, who continued to consult his brother on technical matters and apparently helped support him, reportedly told Preece in the early 1880s that Oliver was becoming too reclusive and suspicious of others to work effectively as part of a group, and he was probably right.[45]

Heaviside's one consolation was his work. The psychological rewards of original research helped to make his meager and solitary life in Camden Town tolerable, even happy. Much later, when he was again in straitened circumstances, he spoke of those years: "There was a time indeed in my life when I was something like old Teufelsdröckh in his garret, and was in some measure satisfied or contented with a mere subsistence. But that was when I was making discoveries. It matters not what others may think of their importance. They were meat and drink and company to me."[46]

Cables and Field Theory

The main focus of Heaviside's work during his years in Camden Town was telegraphic propagation, in particular the distortion that signals suf-

42. Heaviside to FitzGerald, 26 May 1894, FG-RDS; cf. [FitzGerald] 1894. Heaviside mentioned his mother's epilepsy in a letter to Lodge, 11 Jan. 1895, OJL-UCL.
43. Baker 1976: 208; Appleyard 1930: 222, quoting a 22 Nov. 1881 letter to Heaviside from his brother Arthur. It was fortunate for Heaviside that he did not take up the offer of the American job; the English telegraphers who went to help install Wheatstone automatic telegraphs complained of broken promises and mistreatment by Western Union; see U.S. Senate Committee on Education and Labor, 1883 hearings, *Relations between Labor and Capital* 1: 157–58.
44. Heaviside to Highfield [draft], 14 Mar. 1922, OH-IEE.
45. See Baker 1976: 208.
46. Heaviside to Larmor, 18 July 1908, JL-RS. "Teufelsdröckh" was the German "philosopher of clothes" in Thomas Carlyle's *Sartor Resartus* (1838).

fered in passing along submarine cables. The subject had already had a great influence on the evolution of British electrical theory, and it was to lead Heaviside to some of his most important discoveries.

Distortion had first come to the attention of British electricians in 1853 when Latimer Clark of the Electric and International Telegraph Company noticed that signals sent through the new Anglo-Dutch cable were not as sharp and distinct as those on ordinary telegraph lines. He went on to show that pulses sent through long submarine cables were slightly delayed and greatly elongated, so that dots and dashes sent in rapid succession ran together and became unreadable at the receiving end. A similar effect was noticed a short time later on the London–Manchester underground lines owned by the Magnetic Telegraph Company.[47]

Most telegraph engineers regarded such "retardation" as just another of the many obstacles to rapid signaling. But Michael Faraday, to whom Clark demonstrated the phenomenon in October 1853, saw in it a key to some of the fundamental principles of electromagnetism and a means by which he could build support for his own rather unorthodox field ideas.[48] According to Faraday, an electric charge was not an accumulation of an imponderable fluid that exerted forces directly at a distance, but simply a manifestation on the surface of a conductor of a state of strain in the surrounding dielectric; an electric current was not a real flow of fluid, but simply a consequence of the continual breakdown of strain within the conductor. Conduction was always preceded by the induction of a state of strain in and around the conductor, he said, and it was only after a charge had been induced on the surface of the conductor that the current within it could rise to its full strength.[49]

The process of induction followed by conduction usually happened too quickly to be noticed, according to Faraday; ordinary laboratory conductors and even long overhead telegraph lines had such low capacitance that they became fully charged and began to carry a steady current almost instantaneously. But a cable, consisting as it did of a central wire separated from an outer conductor of water or damp soil by only a thin layer of gutta-percha insulation, was in effect a huge condenser or Leyden jar; its capacitance was enormous. When a battery was touched to the central wire of a cable, it took an appreciable time for the electrostatic strain to spread laterally through the gutta-percha along the entire length of the cable, so that the rise of the current was retarded. When the battery was disconnected, it took additional time for the strain

47. Williams 1965: 484–85; Bright 1898: 25.
48. Faraday 1855, 3:508–20 [1854]; see also Hunt forthcoming.
49. Williams 1965: 375.

to relax, and the consequent discharge along the cable greatly prolonged the fall of the current and produced the stretching and blurring of pulses which Clark had observed. The existing action-at-a-distance theories had difficulty accounting for these effects; they indicated that induction should occur at once along the entire cable and made no provision for the observed influence of the gutta-percha dielectric. That it took time for inductive effects to propagate suggested that they acted through a medium, and in the account of cable phenomena that he gave at the Royal Institution early in 1854, Faraday pointed to retardation along cables as "strong confirmation" of his own views on the nature of electrical actions.[50]

The work of Clark and Faraday stirred concern that retardation might make signaling through a proposed Atlantic cable too slow to be profitable, and it was to address this question that William Thomson first took up the theory of telegraphic propagation. In two letters he wrote to G. G. Stokes in October 1854 and published the following May in a paper "On the Theory of the Electric Telegraph," Thomson combined Faraday's idea of lateral induction with Fourier's equations for the propagation of heat to obtain formulas describing the diffusion of current and potential along a submarine cable.[51] Heaviside later said that Thomson's theory marked "the first step towards getting out of the wire into the dielectric," and though it left many steps still to be taken, it set in train a fundamental shift in the way British physicists and engineers conceived of telegraphic propagation.[52]

According to Thomson's theory, retardation increased with both the resistance and the capacitance of the cable and so was proportional to the square of its length. This "law of squares" indicated that the retardation that had already become troublesome on cables a few hundred miles long would be many times worse on the two thousand–mile Atlantic cable; unless expensive measures were taken to reduce its resistance and capacitance, the cable might be limited to one or two words per minute. Special instruments and signaling techniques might increase this somewhat, Thomson said, but there appeared to be no way to escape retardation completely on long cables.

Thomson's conclusions were challenged in 1856 by E. O. Wildman

50. Faraday 1855, 3:508 [1854].
51. William Thomson 1882–1911, 2:61–76 [1855]. Cable telegraphy and retardation attracted great interest in 1854; see several papers in the 1854 *BA Report*, esp. those by C. F. Varley (pp. 17–18) and Frederick Bakewell (p. 147). The account of Thomson's cable theory in Smith and Wise 1989: 446–53 seems to me to overstate the differences between Faraday and Thomson on the nature of induction and retardation.
52. Heaviside 1892, 2:79 [1886].

Whitehouse, a former Brighton surgeon who had become the electrical advisor to the new Atlantic Telegraph Company. He claimed to have found experimentally that the law of squares was invalid and that a cable of relatively small diameter worked at a high voltage afforded the best signaling.[53] The first Atlantic cable was built to Whitehouse's specifications, no doubt in part because this was the cheaper alternative, and after it was successfully laid in August 1858, its operation was put in Whitehouse's hands. His apparatus proved ineffective, however, and the high voltages he used damaged the insulation of the cable, which had already been weakened by improper handling. After two weeks of disputes and confusion, Whitehouse was dismissed, and Thomson, who as a director of the company had taken part in the laying of the cable, was put in charge. His own delicate low-voltage instruments were used with some success, but the damage had already been done, and the insulation gave way altogether by October.[54]

The demise of the Atlantic cable, followed within a year by the expensive failure of the first Red Sea cable, prompted the British government to launch an inquiry into the problems of submarine telegraphy.[55] In its report, the committee largely endorsed Thomson's theory of cable propagation and his techniques for signaling. The new Atlantic cable of 1866 was built and operated in accordance with Thomson's advice, and its great success served to bolster his theory of propagation and with it Faraday's approach to induction phenomena.

Perhaps the clearest and most sweeping statement of the debt that science owed to submarine telegraphy was that Thomson gave in his 1871 presidential address to the British Association. After noting the contribution the work of the Electrical Standards Committee had made to Maxwell's electromagnetic theory of light, Thomson said:

> This leads me to remark how much science, even in its most lofty speculations, gains in return for benefits conferred by its application to promote the social and material welfare of man. Those who perilled and lost their money in the original Atlantic Telegraph were impelled and supported by a sense of the grandeur of their enterprise, and of the world-wide benefits which must flow from its success; they were at the same time not unmoved

53. Whitehouse 1856. This claim led to an exchange between Thomson and Whitehouse in the pages of the *Athenaeum* in Oct. and Nov. 1856; see Thompson 1910, 1:330–32, and Smith and Wise 1989: 661–67. Faraday endorsed Whitehouse's thin cable, apparently not realizing that the advantages of its low capacitance would be outweighed by the disadvantages of its high resistance; see Hunt forthcoming and Bright 1898: 54n.

54. Thompson 1910, 1:366–96; Smith and Wise 1989: 670–75.

55. *Report of the Joint Committee Appointed by the Board of Trade and the Atlantic Telegraph Company to Inquire into the Construction of Submarine Telegraph Cables*, British Parliamentary Papers, 1860 [2744]; cf. Bright 1898: 59–61 and Smith and Wise 1989: 674–78.

by the beauty of the scientific problem directly presented to them; but they little thought that it was to be immediately, through their work, that the scientific world was to be instructed in a long-neglected and discredited fundamental electric discovery of Faraday's, or that, again, when the assistance of the British Association was invoked to supply their electricians with methods for absolute measurement (which they found necessary to secure the best economical return for their expenditure, and to obviate and detect those faults in their electric material which had led to disaster), they were laying the foundation for accurate electric measurement in every scientific laboratory in the world, and initiating a train of investigation which now sends up branches into the loftiest regions and subtlest ether of natural philosophy.[56]

By focusing attention on retardation and other phenomena that fit well with Faraday's "long-neglected and discredited" ideas about the role of the dielectric, submarine telegraphy helped steer the thinking of British scientists and engineers toward acceptance of the field approach to electromagnetism. Submarine telegraphy did not *produce* Faraday's field ideas, but it provided a market for them—a market that, like the cable industry, was almost entirely British. Electricians in Germany, France, and America worked almost exclusively with overhead landlines and rarely encountered retardation effects; only in Britain was there sustained experience with cables and retardation, and only in Britain were Faraday's ideas taken up and elaborated.[57]

Heaviside on Propagation

Heaviside's own concern with electromagnetic propagation grew quite directly out of his experiences on the Anglo-Danish cable. One of his jobs at Newcastle was to help maintain and operate the Wheatstone automatic telegraphs, complex machines that could read messages coded on punched paper tape and transmit them at rates well over one hundred words per minute. They worked so well on land lines that by the 1870s they were used for the bulk of the traffic on Britain's busiest intercity routes, but on cables they were plagued by retardation prob-

56. William Thomson 1889–94, 2:161–62 [1871].
57. Retardation had in fact first been observed by Werner Siemens on underground lines in Germany in 1849 and had led him to views similar to Faraday's on the action of the field. Wilhelm Weber and other German physicists did not accept Siemens's theory, and the failure of the underground lines because of poor construction soon removed the problem from their attention; see Siemens 1966: 75, 129, 164 and Faraday 1855, 3:521 [1854]. Thus while German physicists were familiar with Faraday's writings (see Jungnickel and McCormmach 1986, 1:120), they had relatively little incentive to take up his ideas.

lems.[58] Heaviside found that signals sent through the Anglo-Danish cable became unreadable on ordinary receivers when sent at more than twenty-five or thirty words per minute and that even with a delicate siphon recorder the limit was no more than about seventy words per minute. At that speed, he said, the dashes appeared as large humps and the dots "were mostly wiped out, though it could be seen where they ought to have been, so that some reading was possible."[59]

Heaviside wanted to trace the causes of this distortion, and "The only way, so far as I am aware," he said in 1874, "is to follow the method given by Sir William Thomson in 1855."[60] Teaching himself the necessary mathematics as he went along, Heaviside produced a series of papers in the 1870s in which he applied Thomson's propagation theory to phenomena he had encountered on the Anglo-Danish cable. In "On Telegraphic Signalling with Condensers" (1874), "On the Speed of Signalling through Heterogeneous Telegraph Circuits" (1877), and "On the Theory of Faults in Cables" (1879), he showed how various circumstances could alter the propagation conditions contemplated in the original theory and either increase or decrease the amount of retardation.[61] It was well known, for example, that a leak in the insulation of a cable made for faster signaling, a phenomenon Heaviside had seen for himself when the Anglo-Danish cable developed a serious fault in the early 1870s. He and the other operators were able to send a batch of messages through at high speed just before the cable failed altogether, and these signals, though weak, were "exceedingly clear" and in fact "the best ever got at the speed."[62] As Heaviside explained in his "Theory of Faults," the leak had allowed the charge on the cable to escape more quickly, reducing the amplitude of the signals but improving their clarity. Using Thomson's equations and Fourier series, he showed in detail how this accelerated discharge affected the "arrival curves" for the current and voltage and calculated how much leaks of various sizes would decrease the retardation. He even suggested that "artificial faults" be inserted in cables to speed up signaling, though this idea was slow to win acceptance.[63]

The most important of Heaviside's early propagation papers concerned not an application of Thomson's theory to a special case but an extension of the theory itself. In an August 1876 paper "On the Extra

58. Bowers 1975: 179–80. On the use of Wheatstone automatics on the Anglo-Danish cable, see Holmes 1871; Holmes had supervised the laying of the cable in 1868.
59. Heaviside 1893–1912, 1:423 [1893].
60. Heaviside 1892, 1:48 [1874].
61. Ibid., pp. 47–53, 61–70, 71–95 [1874–79].
62. Heaviside 1893–1912, 1:424 [1893].
63. Heaviside 1892, 1:77 [1879].

Current," Heaviside showed that Thomson's theory was incomplete and that some of his results took on a quite different form when self-induction was taken into account.[64] Self-induction is one of the basic properties of electric currents; discovered by Faraday in the 1830s, it shows itself as a tendency for currents to oppose any changes in their strength. The study of self-induction played an important part in the evolution of field theory, particularly the idea that the magnetic field around a current-carrying wire is a storehouse of energy and motion. Victorian physicists often pictured this field as filled with some sort of whirling machinery in the ether; when a current was started or reversed, it had to react against the momentum of this whirling machinery or, in electrical terms, against the self-induction. Self-induction thus acted much like a flywheel to oppose any change in the current.

Thomson had been able safely to ignore self-induction in his original theory of cable propagation because the self-inductive effect was swamped by the large retardation caused by electrostatic capacitance. He thus found that current and voltage diffused along the cable in much the same way that heat diffuses along a rod. But Heaviside saw that a fully general theory of electromagnetic propagation would have to take self-induction into account. From the basic equations connecting charge, current, and potential, he derived the fundamental "equation of telegraphy," sometimes called "Heaviside's equation":

$$d^2v/dx^2 = ckdv/dt + scd^2v/dt^2.$$

Here v is the voltage at a point x on the line, and c is the capacitance, k the resistance, and s the inductance, all per unit length. Thomson had simply taken $s = 0$.

Using Fourier series, Heaviside showed that his equation implied that when self-induction came into play, the current no longer simply diffused along the line, as in Thomson's theory, but instead oscillated back and forth before settling into a steady state. As Heaviside summed up, "We may be sure that, in virtue of the property of the electric current which Professor Maxwell terms its 'electromagnetic momentum,' whenever any sudden change of current or of charge takes place in a circuit possessing an appreciable amount of self-induction, the new state of equilibrium is arrived at through a series of oscillations in the strength of the current, which may be noticeable under certain circumstances."[65] This was the beginning of Heaviside's reconceptualization of tele-

---

64. Ibid., pp. 53–61 [1876]; cf. Whittaker 1929: 207.
65. Heaviside 1892, 1:59 [1876].

graphic propagation in terms of electromagnetic waves, a step that was to have far-reaching consequences for both theory and practice.

In his early papers on propagation, Heaviside always worked directly in terms of the line parameters: the resistance, capacitance, and inductance of the telegraph line and its terminal apparatus. Some of these papers were mathematically very demanding; even Heaviside described "On Induction between Parallel Wires," published in the *Journal of the Society of Telegraph Engineers* in 1881, as "a very heavy paper."[66] But mathematical sophistication and theoretical depth are not always the same thing, and in the early 1880s, Heaviside realized that to progress further he would have to step back from the complicated equations connecting the line parameters and turn instead to an examination of the fundamentals of electromagnetic theory. His work on signal propagation had led him to focus on field phenomena and had convinced him that there was far more of interest going on in the space surrounding the electrical conductors than in the wires, coils, and condensers themselves. This realization was reflected both in his move in the early 1880s away from the line parameters and toward the electric and magnetic field strengths as the primary quantities of investigation, and in his increasing reliance on Maxwell's theory. But while his methods changed, Heaviside's aim remained the same: he continued, now through the study of fundamental electromagnetic theory, to seek tools that he could bring to bear on the improvement of telegraphy.

Turning to Maxwell

Heaviside came across a copy of Maxwell's *Treatise* soon after it was published in 1873, and he was immediately impressed, he later said, by its "prodigious possibilities," though much of its mathematics was then far above his head.[67] He spent the next several years trying to master the book, and it was not until 1876, in the passage from "On the Extra Current" quoted above, that he first cited it in print. Like many who read the *Treatise* in the 1870s, Heaviside did not at first see exactly how it differed from the earlier work of Faraday and Thomson; in particular, it took him some time to appreciate the importance of Maxwell's new ideas about displacement currents and electromagnetic waves. In a letter to Hertz in 1889, Heaviside remarked that he had by then become so famil-

66. Ibid., pp. 116–41 [1881]; Heaviside notebook 3a, entry 27, OH-IEE.
67. Heaviside to Joseph Bethenode, 24 Feb. 1918, quoted (after retrans.) in Nahin 1988: 24–25.

iar with the theory of electromagnetic waves that he took the results of
Hertz's experiments "almost as a matter of course." But this understanding of the full implications of Maxwell's theory had come slowly. "Take a
suppositious case," he said:

> Supposing, after my first reading of Maxwell, when I imagined I had pret-
> ty well taken it all in, I had gone to sleep for many years; and, on waking,
> had had your experiments to read about, and the doctrine in explanation
> thereof. It would have been a perfect revelation to me! So I think it must be
> even to many readers of Maxwell, who have read, but have done no more;
> in particular, have not sat down and developed that chapter of his on Prop-
> agation into its consequences.[68]

It was by changing from a reader of the *Treatise* to a student of it, in
particular by working out its consequences and applying it to problems
of telegraphic propagation, that Heaviside acquired his remarkable
grasp of Maxwell's theory.

Heaviside said later that he had been "expounding Maxwell since
1882," and that seems an appropriate year from which to date his emergence as a public exponent of Maxwell's theory.[69] He referred to the
transition in a footnote to "The Theory of Propagation of Current in
Wires," a paper he wrote in 1882 but did not publish until ten years later.
It completed his initial series on cable signaling, he said, and formed "a
sort of missing link between the earlier articles on propagation and the
later ones, in which the subject is discussed on the basis of Maxwell's
theory of the ether as a dielectric."[70] Virtually everything Heaviside
published after 1882 concerned the elaboration and application of Maxwell's theory.

At the same time he was changing his approach to electromagnetic
theory, Heaviside was moving into a new forum for expressing his views.
Toward the end of 1882 he began to publish regularly in the *Electrician*, a
weekly London trade and technical journal. Except for a gap in the late
1880s, his writings appeared there almost continuously until 1902; in all,
he published enough in the journal to fill about seventeen hundred
pages in his collected papers.[71] The *Electrician* was a commercial journal
aimed mainly at engineers and businessmen in the rapidly growing electrical industries, and much of its space was filled with "Trade Notices"
and reports of lawsuits and business maneuverings. But its proprietor,

68. Heaviside to Hertz, 13 July 1889, HH-DM.
69. Ibid., 14 Feb.
70. Heaviside 1892, 1:141n [1882]; cf. p. vii.
71. About half the 1,100-page *Electrical Papers* and three-quarters of the 1,500-page
Electromagnetic Theory originally appeared in the *Electrician*.

cable magnate Sir John Pender, professed to want the journal to aspire to "higher things" and to serve the diffusion of electrical knowledge; no doubt he also hoped it would help improve the public image of his companies, whose virtual monopoly of the world cable network made them the target of suspicion and animosity.[72] Under C. H. W. Biggs, editor from its foundation in 1878, the *Electrician* published extensive reports of scientific meetings, accounts of inventions and discoveries, articles instructing readers in electrical principles, and even discussions of quite arcane theoretical points.

Heaviside's work first appeared in the *Electrician* in 1879, when he sent it his article "Sensitiveness of Wheatstone's Bridge," which *Nature* had turned down.[73] But he did not publish there regularly until November 1882, when, at Biggs's request, he began his series of articles "The Relations between Magnetic Force and Electric Current." He tried in these to give an elementary treatment of what he called the "higher conceptions" of electromagnetism as developed by mathematical physicists "from Ampère down to Maxwell," with a strong emphasis on the views of the latter. Although he admitted that these conceptions were "usually supposed to be within the reach of none but mathematicians," Heaviside claimed they "could be to a great extent stripped of their usual symbolical dress, and in their naked simplicity made to appeal to the sympathies of the many."[74]

Heaviside's professed aim in these articles and in most of his later writings in the *Electrician* was to teach, and he tried in them to present electromagnetic principles as clearly and simply as he could. He was not always successful and in fact became legendary for the opacity of his mathematical writings; as FitzGerald later remarked, in Heaviside's "most deliberate attempts at being elementary, he jumps deep double fences and introduces short-cut expressions that are woeful stumbling blocks to the slow-paced mind of the average man."[75] Ironically, one of the main stumbling blocks was Heaviside's use of vectors. Vector analysis is now the universal language of electromagnetic theory, not least through the efforts of Heaviside himself; but in the 1880s it seemed foreign and forbidding to most physicists and electrical engineers. Even Lodge, while praising Heaviside's mastery of electrical theory, described his papers as difficult, "eccentric, and in some respects repellant."[76] But

72. See Pender's remarks at the 24 Jan. 1889 meeting of the Eastern Telegraph Co., in *Elec.* 22 (1889): 382, and the editorial "Ourselves," *Elec.* 10 (1883): 517.
73. Heaviside 1892, 1:8–13 [1879]; Heaviside notebook 3a, entry 29, OH-IEE.
74. Heaviside 1892, 1:195 [1882]; see also Biggs to Heaviside, 1 Sept. 1882, OH-IEE.
75. FitzGerald 1902: 293 [1893].
76. Lodge 1888c: 236.

in his 1882 articles Heaviside managed to restrain his tendency to run on ahead of his readers and produced a reasonably clear introduction to the main points of field theory. Its centerpiece was a discussion of how the vector operators "divergence" and "curl" could be used to derive the strengths of charges and currents from a knowledge of the forces and fluxes in the surrounding field. Heaviside's experience with cables had made him a strong supporter of Faraday and Maxwell's doctrine that the field was the real seat of electromagnetic phenomena, and he sought in his *Electrician* articles to provide mathematical and conceptual tools suited to this approach.

Biggs was pleased with Heaviside's articles. They were, he said, "just what I want": elementary but not simplistic accounts of electromagnetic theory. In December 1882 he asked Heaviside to continue, telling him, "As long as you remain good to write so long shall I be pleased to receive and insert your MS.—I may say that I hope this will be a period of indefinitely long duration."[77] Heaviside took this offer seriously, and it became a turning point in his life. For the first time since leaving Newcastle, he had a job, of a sort—the *Electrician* paid him about forty pounds a year for his articles. This was very meager, of course; Heaviside himself said that he "earned less than a hodman."[78] But his needs were modest, and forty pounds apparently sufficed. More important than the money was the fact that Biggs had given him a steady outlet for his writings. Working full time on his research, Heaviside was producing masses of material and faced, as he later said, "the impossibility of getting rid of it all at once."[79] The *Electrician* was able to absorb far more than any other publication open to him (an average of two or three articles every month from 1883 to 1887), and though it was not an ideal outlet for some of his more advanced work—especially since it did not always come to the attention of foreign scientists, nor even of many in England—Heaviside's relationship with Biggs's journal was, on the whole, very fruitful.[80]

That Heaviside's increasingly sophisticated papers on Maxwell's theory first appeared in an electrical trade journal, squeezed between advertisements for batteries and copper wire, was not really as incongruous as

---

77. Biggs to Heaviside, 5 Dec. 1882, OH-IEE.
78. Heaviside to FitzGerald, 13 Feb. 1894, FG-RDS. On Heaviside's income in this period, see Heaviside to Lodge, 20 Apr. 1892, OJL-UCL, and John Perry to FitzGerald, 3 Feb. 1894, FG-RDS. The average family income for an unskilled laborer in England in 1867 was between £45 and £50 a year.
79. Heaviside to Hertz, 14 Feb. 1889, HH-DM.
80. The *Electrician* was not available to Hertz in Karlsruhe (see Hertz 1893: 196n), and Rayleigh did not see Heaviside's results until they appeared in revised form in the *Phil. Mag.*; see Lodge to Heaviside [draft], 23 Sept. 1888, OJL-UCL.

it might now appear. His work in electromagnetic theory was, like the *Electrician* itself, largely a byproduct of the growth of the British cable industry. Submarine telegraphy had played an important part in the early spread of field ideas in Britain, and it provided the context for virtually all of Heaviside's electrical studies, from his initial investigations of retardation on the Anglo-Danish cable to his most highly developed analyses of electromagnetic propagation. He continued to draw research problems from telegraphy throughout his career, and when in the late 1880s he began to come into direct contact with other students of Maxwell's theory and to take his place as one of the acknowledged authorities on the subject, Heaviside's contributions to the emerging Maxwellian synthesis continued to be shaped in deep and pervasive ways by his roots in telegraphy.

CHAPTER FOUR

Ether Models and
the Vortex Sponge

The Maxwellians were not content to confine their attention to the purely electromagnetic aspects of Maxwell's theory. Like a long line of earlier British physicists, including William Thomson, G. G. Stokes, and Maxwell himself, they regarded all physical phenomena as essentially mechanical, and they sought to explain the electromagnetic equations in terms of the structure and motions of an underlying ether. This mechanical program reached its zenith in Britain in the 1880s and 1890s, and its adherents came to view pure matter and motion as the only truly explanatory basis for any physical theory. The idea of direct action at a distance struck most of them as absurd; by the last quarter of the nineteenth century, the electric and magnetic fluids were, in their view, "played out" and "fast evaporating into nothingness."[1] Heaviside spoke for a generation of his colleagues when he declared, in 1888, "Concerning the often-asked question, What is electricity? I can attach but little importance to the answer by itself. But," he said, "the question, What is the mechanism of electrical phenomena? is quite another thing."[2] The discovery, or perhaps invention, of that mechanism became a major goal of the Maxwellian program.

Victorian physicists' pursuit of the mechanism of phenomena was closely tied to their use of a characteristic tool: the mechanical model. Some of their models were simply "illustrations," intended as aids to the

---

1. Heaviside 1892, 1:295 [1883]; cf. FitzGerald to Heaviside, 26 Sept. 1892, OH-IEE, saying that a recent French theory was a return "to the old vomit of action at a distance."
2. Heaviside 1892, 2:486 [1888].

learner; others, notably the vortex sponge FitzGerald devised in 1885, were meant to be actual "likenesses" of the ether. Still others, such as Maxwell's vortex model and FitzGerald's "wheels and bands," were used as genuine research tools and helped lead to solid scientific advances, particularly in the understanding of displacement currents and electromagnetic waves. Even when the more explicit mechanical models began to fade from the scene toward the end of the century, the British preference for vivid representations continued to show itself in the development of "mathematical models" of physical phenomena. Heaviside in particular deliberately constructed his vector system so that it would represent the electromagnetic field as closely as possible and used it to display the workings of the field as clearly and almost as concretely as could any mechanical model.

Models

Maxwell favored the use of models in part as a way to broaden the range of acceptable paths to truth. "For the sake of persons of . . . different types, scientific truth should be presented in different forms," he said in an 1870 address the British Association, "and should be regarded as equally scientific, whether it appears in the robust form and the vivid colouring of a physical illustration, or in the tenuity and paleness of a symbolical expression."[3] He had in mind the work of Faraday, whose ideas had been ignored or attacked by mathematical physicists because they were expressed in images rather than equations. Maxwell had first elaborated his method of physical analogies in the 1850s, as a way to bridge the gap between Faraday and the mathematicians by showing that their concepts were intertranslatable. Although Maxwell was himself an outstanding mathematical physicist, he thought it a mistake to regard physics as exclusively mathematical and to exclude those who thought in terms of physical images rather than equations. Such men, Maxwell said, not only picture the workings of phenomena to themselves; they

> are not content unless they can project their whole physical energies into the scene which they conjure up. They learn at what rate the planets rush through space, and they experience a delightful feeling of exhilaration. They calculate the forces with which the heavenly bodies pull at one another, and they feel their own muscles straining with the effort. To such men momentum, energy, mass, are not mere abstract expressions of the

3. Maxwell 1890, 2:220 [1870].

results of scientific enquiry. They are words of power, which stir their souls like the memories of childhood.[4]

As Maxwell's tone makes clear, he was himself one of those for whom physics was an intensely vivid and tactile pursuit. He was far from alone; Lodge, for one, said that on reading Maxwell's remarks, "I remember feeling enthusiastically that that was exactly my own case."[5] It was men of this type, with their strong physical imaginations and their reliance on concrete models and analogies, who gave Victorian physics much of its distinctive character.

One of the most important uses Victorian physicists made of models was pedagogical. To help students understand an unfamiliar phenomenon, such as the charging of a condenser, a physicist would devise a mechanical "illustration" of it in which the relations of the corresponding parts could more easily be traced. William Thomson often used models in this way "to fix ideas," and Maxwell declared that illustrative models not only were "convenient for teaching science in a pleasant and easy manner" but provided the best way to translate complex mathematical relationships into a concrete and readily grasped form without a loss of rigor. Indeed, he said, "If science is ever to become popular, and yet to remain scientific," it must be by the use of physical illustrations; for these provided the only means by which many areas of physics could be presented in a rigorous yet understandable way to those for whom a bare set of equations would be wholly opaque.[6]

Maxwell sometimes built working versions of his models, such as the gear-and-crank device he used at the Cavendish Laboratory to illustrate mechanically the induction of electric currents.[7] Most of his models, however, were purely imaginary, their workings meant to be followed only in the mind's eye. The "Mechanical Illustration of the Properties of a Dielectric" Maxwell described in his *Treatise* existed only on paper, but this imaginary set of tubes and pistons helped him to explicate in an easily followed way the fundamental phenomenon of Maxwellian electrodynamics, the charging of a condenser.[8] A solid grasp of the concept of displacement was crucial to any real understanding of Maxwell's theory, and his model dielectric illustrated the idea more clearly and vividly than pages of ordinary exposition could. Maxwell's successors shared his

4. Ibid.
5. Lodge 1931b: 21.
6. Maxwell 1890, 2:219 [1870].
7. Campbell and Garnett 1882: 551–53. This device is now at the Whipple Science Museum in Cambridge.
8. Maxwell 1873, art. 334.

belief in the pedagogical value of models, especially in illustrating the concept of displacement. Heaviside used a simplified piston model of a condenser in his own writings; Lodge relied heavily on the "string and button" and "hydraulic Leyden jar" models he devised to illustrate displacement phenomena.[9]

These illustrative models were meant to be instructive, not explanatory. They added nothing essentially new to the theoretical account of a phenomenon but simply made it more vivid. They were certainly not meant to be realistic: Maxwell did not think electromagnetic induction was traceable to the turning of cranks and the meshing of gears, nor did Heaviside and Lodge think dielectric polarization was really a matter of movable pistons or of strings and buttons. As FitzGerald put it, illustrative models were intended as *analogies* of phenomena, not *likenesses;* they offered a similitude of relations, not of things.[10]

Some models, however, were regarded as more than mere illustrations. Because most Victorian physicists believed matter and ether were fundamentally mechanical, they could hope that a carefully chosen mechanical model might truly depict the underlying physical reality, or at least some aspects of it; a model might come close to being a *likeness* of nature. Moreover, by the last third of the nineteenth century, they had before them a strong example of such a successful and realistic model in what Heaviside called "that remarkable triumph of hard-headed men, the kinetic theory of gases." Here, "on the ridiculously simple hypothesis that a gas consists of an immense number of small particles in motion," colliding randomly with one another in accordance with the laws of chance and mechanics, was a theory that accounted in a remarkably full and detailed way for most of the known properties of gases and even predicted new ones that were confirmed by experiment. This model, Heaviside said, "was something deserving the name of explanation," and his colleagues agreed; by the 1870s, virtually all British physicists were convinced that the kinetic model reflected the true nature of gases.[11] There might be imperfections in the theory and points requiring further study, but it seemed clear that its basic features were correct and that the image of tiny particles colliding with one another constituted a real likeness of nature.

It ought to be possible, at least in principle, to do the same thing for the ether: to find a mechanical model that reflected its true nature. So, at least, many British physicists believed; and in the latter half of the nine-

9. Heaviside 1892, 1:479–80 [1885]; Lodge 1889b: 32–62.
10. FitzGerald 1902: 167; from an unfinished essay "Foundations of Physical Theory: Function of Models" written 1885–88 but first pub. 1902.
11. Heaviside 1892, 1:336 [1884].

teenth century their faith in the explanatory power of mechanisms led them to devise a profusion of mechanical ether models during what Martin Klein has aptly called "the High Baroque phase of the mechanical world view."[12] As they pushed the method of mechanical explanation to its limits, they produced some of the most subtle and ingenious examples of physical speculation ever devised.

The most fruitful of these ether models, and the most important for the development of electromagnetic theory, was the vortex and idle-wheel mechanism Maxwell described in his 1861–62 paper "On Physical Lines of Force."[13] This model led Maxwell to some of his most important ideas, including displacement currents and the electromagnetic theory of light, and it had a strong and lasting influence on later students of his theory. J. J. Thomson recalled in 1931 that on reading Maxwell's paper as an undergraduate, he was "raised to such a pitch of enthusiasm . . . that I copied out the whole of it long-hand, and it is a very long paper." Both Lodge and FitzGerald devised their own variants of Maxwell's model, while Heaviside praised it as an example of "the highest kind of scientific speculation" and declared that it was sure to prove "very useful in its suggestiveness to future electrical students."[14]

Maxwell's model functioned on several levels, and while some parts of it were intended as no more than convenient hypotheses, others were meant to represent the real structure of the ether. Maxwell believed the molecular vortices to be almost certainly real; citing William Thomson's analysis of the Faraday effect, he declared in his *Treatise* that "we have good evidence for the opinion that some phenomenon of rotation is going on in the magnetic field" and that "this rotation is performed by a great number of very small portions of matter, each rotating on its own axis."[15] The idea of idle-wheel particles rolling between the vortices, on the other hand, he regarded as no more than a contrived and "somewhat awkward" hypothesis. "I do not bring it forward as a mode of connexion existing in nature," he said in "Physical Lines," but merely as one that was easily investigated; and since many of his results were independent of the details of how the vortices were connected, he thought it best to pick a particular mechanism and follow out its consequences.[16] As he told P. G. Tait in 1867, the purpose of the vortex and idle-wheel model was

12. Klein 1972: 73.

13. Maxwell 1890, 1:451–513 [1861–62]; see Siegel 1986 for a thorough analysis of Maxwell's model.

14. J. J. Thomson 1931: 34; Lodge 1889b: 177–216, 263–64; Heaviside 1892, 1:333–34 [1884].

15. Maxwell 1873, art. 831.

16. Maxwell 1890, 1:486 [1861].

"to show that the phenomena are such as can be explained by mechanism. The nature of the mechanism is to the true mechanism what an orrery is to the Solar System." Maxwell continued to search for the "true mechanism" of the ether, but in the meantime he thought it best to put his main results on a broader and less hypothetical basis. The "Dynamical Theory of the Electromagnetic Field" he published in 1864 was, as he told Tait, "built on Lagrange's Dynamical equations" and was "not wise about vortices."[17] Maxwell himself continued to be "wise about vortices," however, and when in his *Treatise* he returned to the Faraday effect (a problem he had left aside in "Dynamical Theory"), he again relied heavily on the vortex model.

Maxwell's writings, particularly "Physical Lines" and "Dynamical Theory," reflect the two parallel endeavors in which he was engaged. On the one hand, he sought to formulate the complete macroscopic laws of electromagnetism, the "field equations." On the other, he sought "the true mechanism" of the ether by which the laws of electromagnetism would find a deeper, even ultimate, explanation in terms of matter and motion. There was sometimes a tension between these two aims, and Maxwell oscillated between focusing on one, as in "Physical Lines," and the other, as in "Dynamical Theory." But in fact the two endeavors were complementary, each serving to illuminate and advance the other: study of the laws and phenomena of electromagnetism helped suggest mechanisms in the ether, as analysis of the magnetic rotation of light had suggested the existence of vortices, while the exploration of these mechanisms could lead in turn to new laws, such as those governing displacement currents and electromagnetic waves.

Both the search for laws and the search for mechanisms were continuing themes in the work of Maxwell's followers. While the Maxwellians were exploring, extending, and revising Maxwell's field equations, they also continued to devise mechanical models, both as a step toward finding the true mechanism of the ether and as a way to make the meaning of the electromagnetic laws clearer and more vivid.

Wheels and Bands

The most important of the Maxwellians' ether models was the "wheel and band" model FitzGerald devised in 1885. It was similar to Maxwell's vortex model but mechanically much simpler. It was simple enough, in fact, for FitzGerald to have a working version of it built—"rather pretty

17. Maxwell to Tait, 23 Dec. 1867, in Knott 1911: 215.

4.1. FitzGerald's wheel-and-band model of the electromagnetic ether (unstrained). Each wheel is connected by rubber bands to its four neighbors; all can spin freely together without straining the bands. This spinning represents a steady magnetic field.

on a mahogany board with bright brass wheels," he told Lodge—and to use it to illustrate Maxwellian ideas to lecture audiences.[18] But the model was not just a pedagogical aid; it also served as a research tool and helped in important ways to guide FitzGerald's thinking about electromagnetism.

FitzGerald hit upon his wheel-and-band model at the very end of 1884 while looking for a way to depict mechanically the flow of energy in an electromagnetic field. J. H. Poynting's discovery of the paths of this flow about a year before is discussed more fully in the next chapter; it had a profound effect on the way Maxwell's theory was understood, not least through the model it inspired FitzGerald to devise. FitzGerald immediately sent an account of his model to Lodge, who had already invented several ether models of his own. "I have been constructing a model ether," he told Lodge, "and if it is the same as yours I want to give you credit for it, as I want it to illustrate Poynting's great discovery that the energy of an electric current must come in at its sides and is not carried along with the current."[19]

FitzGerald's model was very simple. "I propose," he told Lodge, "a series of wheels of course on fixed axes all connected in pairs by india-rubber bands" (see Figure 4.1). The spinning of the wheels represented a magnetic field, much as in Maxwell's vortex model; their rotational

18. FitzGerald to Lodge, 3 Mar. 1894, OJL-UCL. E. T. S. Walton recalls having seen the model at Trinity College Dublin as late as the 1960s, but a search of the attics and storerooms of the Physics Building in 1982 failed to locate it. Inquiries by Denis Weaire indicate that it was probably thrown out during a general housecleaning in the early 1970s.

19. FitzGerald to Lodge, 1 Jan. 1885, OJL-UCL. On the workings of the model, see FitzGerald 1902: 142–62 [1885] and Hunt 1987. FitzGerald spoke "On the Transference of Energy in the Electro-magnetic Field, Prof. Poynting" at the Royal Dublin Society on 15 Dec. 1884, but only the title was published; see *Nature 31* (1885): 330.

inertia corresponded to self-inductance. The elastic bands served in place of Maxwell's idle wheels to convey motion from one wheel to the next. If all the wheels in a region were set spinning at the same speed, there would be no consequent strain on the bands. But if some wheels were turned through different angles than their neighbors—if, for example, the wheels in one region were held fixed while those in another were turned—there would be a resultant strain on the bands. The elements of the medium would become "polarized," FitzGerald said, with one side of each band tightened and the other loosened. This represented an electric field, with the elasticity of the bands corresponding to the "inductive capacity" of the medium.

The workings of FitzGerald's model can best be grasped by examining how it could be used to depict a particular phenomenon, such as the charging of a condenser; this will also bring out more clearly the distinctive way in which the Maxwellians conceived of electromagnetic phenomena. Imagine two regions from which the bands have been removed; these represent perfect conductors. Then imagine the bands to be removed along a line connecting the two regions; this represents a conducting channel, or wire. If we now turn the wheels above this channel in one direction—say, clockwise—and those below it in the other, the two conductors will become "charged" oppositely (see Figure 4.2). Notice that this "charge" appears purely as a reflection of conditions in the medium, a result characteristic of the Maxwellian view, which focused on the field and regarded changes and currents as derivative phenomena.

Because of the way the wheels are connected to one another throughout the region around the charged conductors, the bands will be strained and will try to turn the wheels back toward their original positions; the charge can be maintained only by continuing to pull along the conducting channel with an "impressed force" (corresponding to a battery or dynamo). But now imagine the bands along the channel to be replaced; this corresponds to insulating the two conductors by removing the connecting wire. The tension in the surrounding bands will now be partly spent in straining the replaced bands, and a self-locked system of strain will be established throughout the region between and around the conductors (see Figure 4.3). Each rubber band will be loose on the side toward the first conductor and tight on the side away from it. A vector drawn from the loose side to the tight side represents the electric "displacement" within each element; the pattern of lines thus generated is the same as that formed by the lines of force around two charged bodies. Notice that nothing is actually "displaced" along such a line of displacement; there is only a rearrangement of tension, or a "change of struc-

4.2. FitzGerald's wheel-and-band model (strained). A battery connected between two conducting plates will charge them oppositely and set up an electrostatic field in the intervening space. This process can be depicted on FitzGerald's model by removing the bands from two regions (the "plates") and along a connecting channel (the "wire") and then straining the surrounding bands. The large arrow represents the "impressed force" of the battery; when applied along the connecting channel, it turns the surrounding wheels in the directions and amounts shown by the curved arrows. Neighboring wheels are turned different amounts, straining the bands connecting them; the narrow lines represent tightened bands and the shaded lines loosened ones. Note that the bands around the left plate are all loose, while those around the right plate are all tight; this condition corresponds to a positive charge on the left plate and a negative charge on the right one. Since the strained bands will try to turn the wheels back to their original positions, the opposite charges of the plates will persist only as long as the impressed force is kept up.

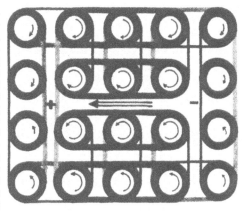

ture," in the elements along it. This was an important point, and FitzGerald later made much of it.

Now consider how FitzGerald's model could depict the discharge of a condenser and the accompanying flow of current and energy. Imagine the bands along a line connecting the two charged regions to be partly loosened, so that they slip slightly over the wheels and dissipate part of their energy in friction. This represents a conducting path, but one with some resistance; low friction corresponds to a good conductor, high friction to a bad one, and the heat generated by the rubbing of band against wheel represents the ohmic heating of a current-carrying wire. With such a conducting path available, the strain in the medium is no longer self-locked; the bands along the conducting path will begin to slip, representing a conduction current, while the wheels on either side will begin to turn in opposite directions, representing the accompanying magnetic field in the surrounding space. The energy stored in the strained bands will gradually be dissipated as heat in the conducting "wire" as the medi-

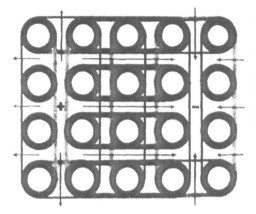

4.3. FitzGerald's wheel-and-band model (strained and locked). If the impressed force in Figure 4.2 is removed and the missing bands along the conducting channel replaced—that is, if the battery wire is cut and the two plates insulated from each other to form a condenser—the result is the self-locked system of strain shown here. With the impressed force removed, the strained bands try to return to their original unstrained state, turning the wheels slightly and stretching the replaced bands until the tensions balance. The plates retain their "charge" and are surrounded by a pattern of strained bands corresponding to the electrostatic field around a charged condenser. The arrows indicate the lines of electric displacement, running from the loose (shaded lines) to the tight (narrow lines) side of each band.

um returns to its original unstrained and motionless state. The energy comes in from the surrounding medium along the length of the strained bands and enters the "wire" at its sides, just as in the Poynting flux that FitzGerald had originally devised his model to illustrate. In the model, as in the electromagnetic field, energy does not flow along the line of the current but instead in paths perpendicular both to the axes of the wheels (the direction of the magnetic force) and to the "polarization" lines from the loose to the tight sides of the bands (the electric displacement vector).

The model could also illustrate other characteristic Maxwellian phenomena, including the generation of electromagnetic waves. If the friction along a conducting path like that described above is not very great—if the "resistance" is low—then, when the impressed force is removed, the inertia of the wheels will carry them past the equilibrium point and the wheels and bands will "bounce" from one state of strain to its opposite in oscillations of gradually decreasing amplitude. Such an oscillatory discharge will cause the wheels in the surrounding field to bounce as well; they will turn slightly from side to side while the connecting rubber bands alternately tighten and loosen, and a wave of such changing motion and strain will spread out from the discharging wire. The motion of the wheels and the changing strain of the bands correspond to oscillating magnetic and electric fields, respectively; together they represent a train of electromagnetic waves spreading out from a discharging condenser in just the way FitzGerald had first described in

1882. FitzGerald even showed how a sudden slippage of the bands would mimic a spark discharge. "Could anything be more complete!" he exclaimed to Lodge. "The whole thing seems a very fair representation of the ether."[20]

Indeed, the representation was more than "very fair"; it was exact. Early in 1885, in his first publication on the subject, FitzGerald showed that "the equations representing the energy of the model are of the same form as those of Maxwell representing the energy of the ether."[21] The potential energy of the strained bands was proportional to the square of their "displacement," just as in Maxwell's term for the electric energy, while the kinetic energy of the spinning wheels was proportional to the square of their rotational velocity, just as the magnetic energy was proportional to the square of the magnetic force. The Hamiltonian function governing the model was identical to that of the electromagnetic field, so that mathematical reasoning could be transferred freely between Maxwell's theory and FitzGerald's model; the model was an accurate embodiment of Maxwell's equations.

Strictly speaking, the correspondence between theory and model was exact only when Maxwell's equations were restricted to two dimensions; the wheels and bands became tangled if one tried to extend them to three dimensions. But FitzGerald devised another model, consisting of paddle wheels pumping fluid between elastic partitions, which, though far too complicated actually to be built, could be shown to yield the full three-dimensional form of Maxwell's energy terms. From this equivalence, FitzGerald pointed out, "It follows at once that all the results deduced from this form of the energy"—including all the phenomena of Maxwellian electrodynamics and of optics—"can be reproduced on a model." Indeed, he said, "It would even be possible to reproduce external and internal conical refraction and the other peculiarities of the wave surface."[22] FitzGerald's point was that his paddle-wheel model could mimic not only the gross features of the electromagnetic field but its subtlest details as well. It thus provided a strong argument that there was nothing in electromagnetic or optical phenomena that was intrinsically beyond the reach of mechanical explanation.

As seriously as FitzGerald took his models, however, he was fully aware of their limitations. They were only *analogies,* he said, and had to be carefully distinguished from *likenesses.* A good model might be "so far

20. FitzGerald to Lodge, 1 Jan. 1885, OJL-UCL.
21. FitzGerald 1885a.
22. Ibid., and FitzGerald 1902: 150 [1885]. In citing conical refraction, FitzGerald was invoking an old Trinity College Dublin tradition: it had been predicted by Sir William Rowan Hamilton and demonstrated by Humphrey Lloyd in 1832.

analogous to [the ether] that we may in many respects reason safely from one to the other," but FitzGerald warned against uncritically accepting such a model as a true depiction of reality. "The danger," he said, "is that we may be satisfied with an analogy, and mistake it for a likeness" and so be led to premature and erroneous conclusions about the nature of reality.[23] "I do not in the least intend to convey the impression," he declared in an account of his models, "that the actual structure of the aether is a bit like what I have described." If the ether could somehow be examined at great magnification, it would certainly not, he said, be found to be "actually made up of wheels and india-rubber bands, nor even of paddle-wheels, with connecting canals."[24] The wheel-and-band and paddle-wheel models were far from being likenesses of the ether; their value, which was considerable, was as analogies.

Changing Displacement

FitzGerald could reason more vividly and concretely about the workings of his wheel-and-band model than about disembodied equations, and he often used it as a guide in his studies of Maxwell's theory. "I . . . am working more or less on my own model ether for concrete ideas," he told Heaviside in 1893 while puzzling over the possible inertia of currents, and the model proved very valuable as such a testing ground for new ideas.[25] One of the strongest immediate effects the model had on FitzGerald's understanding of Maxwell's theory was to reinforce his belief that Maxwell's term *electric displacement* should not be taken literally and that "the word 'displacement' was unfortunately chosen." Maxwell and his early followers had usually treated "displacement" as an actual motion of the ether, an idea that fit with the older elastic solid theories. But FitzGerald found that this literal interpretation led to contradictions, and he thought it "much more likely that what [Maxwell] called 'electric displacements' are changes in structure of the elements of the ether, and not actual displacements of the elements."[26] He used his wheel model to illustrate this point, showing how "electric displacement" in the model arose from a change in the tension of a band rather than from

23. FitzGerald 1902: 166–67 [1885–88].
24. Ibid., pp. 162, 151 [1885].
25. FitzGerald to Heaviside, 21 June 1893, OH-IEE.
26. FitzGerald 1902: 173 [1885]. On Maxwell's literal interpretation of "displacement," see Whittaker 1951: 296n; cf. G. M. Minchin to Lodge, 30 Sept. 1885, OJL-UCL, in which Minchin tells how FitzGerald resolved an apparent contradiction in Maxwell's stress theory of electrostatics by showing that it arose from interpreting "displacement" literally.

its actual displacement. The move away from a literal interpretation of "displacement" FitzGerald was advocating marked a major step in the evolution of Maxwell's theory and helped free it from some of the main inconsistencies of its original formulation.

The question of the true nature of displacement became an especially lively issue early in 1885 in the wake of William Thomson's famous Baltimore lectures on the wave theory of light. In them he had essentially endorsed the elastic solid or "jelly" theory of the ether, with light waves conceived as real bodily vibrations in such a jelly. In a lecture at Philadelphia, he had gone so far as to state flatly that "the luminiferous ether is an elastic solid" and to point to a bowl of jelly as "the nearest analogy [to the ether] I can give you."[27] But the elastic solid had serious shortcomings even as a purely optical ether, being especially plagued by unwanted longitudinal waves, and it failed altogether to account for electromagnetic phenomena. If an elastic solid ether were not to interfere with the free motion of planets and other bodies, it must, Thomson said, be a very soft substance like pitch or wax, able to sustain rapid vibrations but giving way with little resistance before a steady force. But an ether that gave way so easily could not be the medium by which strong and steady electromagnetic forces were exerted; as FitzGerald noted, it was difficult to see "how such a soft material could be the means by which tramcars are driven by shearing stresses."[28] Nor was it easy to see how permanent magnets or steady electric currents could exist in a jelly ether. Thomson's well-known rejection of Maxwell's electromagnetic theory of light in his Baltimore lectures is largely traceable to his faith in the elastic solid ether and his realization that Maxwell's theory was incompatible with it. But where Thomson took this as grounds for rejecting Maxwell's theory, FitzGerald saw it instead as a reason to abandon the elastic solid ether and the literal interpretation of "displacement" that went with it.

FitzGerald brought this new view of the ether into sharp contrast with Thomson's more orthodox position in two letters he wrote to *Nature* in the spring of 1885. When an account of Thomson's Baltimore lectures had first appeared, FitzGerald had thought it must be mistaken; he could not believe that Thomson had misunderstood Maxwell's theory of electromagnetic waves in the way the words quoted seemed to indicate.[29] But Thomson replied that he had indeed been quoted accurately, prompting FitzGerald to draft a longer letter to *Nature* in which he

27. William Thomson 1889–94, 1:334 [1884].
28. FitzGerald 1902: 508 [1900].
29. Forbes 1885; FitzGerald 1885b; see also Smith and Wise 1989: 459–63.

showed in some detail that the supposed waves *in* telegraph wires that Thomson had described were very different from the waves *in the surrounding ether* that were the focus of Maxwell's theory.[30] FitzGerald's real point, however, came in his final paragraph: "I cannot conclude without protesting strongly against Sir William Thomson's speaking of the ether as *like* a jelly. It is in some respects *analogous* to one, but we certainly know a great deal too little about it to say that it is *like* one." The jelly ether was no more than a misleading analogy, he said, and had to be given up before Maxwell's theory could be properly understood.

FitzGerald had great respect for Thomson. "My letter to *Nature*," he told him, "is not intended to inform *you* of anything as I might as well teach my grandmother to suck eggs."[31] But he was convinced that Thomson's continued advocacy of the jelly theory was a serious mistake, and he feared that the acknowledged leader of British physics was "lending his overwhelming authority to a view of the ether which is not justified by our present knowledge, and which may lead to the same unfortunate results in delaying the progress of science as arose from Sir Isaac Newton's equally guarded advocacy of the corpuscular theory of optics."[32] FitzGerald hoped that his wheel-and-band and paddle-wheel models would help physicists keep their minds open to alternatives. "It is worth while considering these models," he said,

> because in them the disturbance which represents light is not the same as the vibrations of an elastic jelly, for what represents an electric displacement is a change of structure of an element, and not a displacement of the element; and it seems almost certain that, notwithstanding the very high authority which seems to support the view that the aether is *like* an elastic jelly, nevertheless its vibrations are much more of the nature of alterations in structure than of displacements.[33]

FitzGerald wanted the true nature of "displacement" to be left an open question; it should, he believed, be settled by further exploration and experiment rather than by appealing to its correlate in a possibly inappropriate model, such as a jelly. It is interesting that he used his own very specific models of the ether to make this point and to show that electromagnetic and optical phenomena could be represented by mechanisms quite different from those of the orthodox elastic solid theories.

30. FitzGerald 1902: 170–73 [1885].
31. FitzGerald to William Thomson, 25 Apr. 1885, WT-ULC. See Smith and Wise 1989: 459–62, 492–94 on FitzGerald's long and mostly unsuccessful campaign to win Thomson over to Maxwellianism.
32. FitzGerald 1902: 173 [1885].
33. Ibid., p. 162 [1885].

FitzGerald's models played an important part in his campaign in the 1880s to change the way "displacement" was interpreted in Maxwell's theory and to complete the break between the Maxwellian electromagnetic ether and the old jelly theories.

FitzGerald's exploration of the workings of his wheel-and-band and paddle-wheel models marked perhaps the highest stage of the British model-building tradition. But if in FitzGerald that tradition reached the "High Baroque," in the work of his friend Oliver Lodge it began to verge on the rococo. Lodge's enthusiasm for ingenious mechanisms sometimes led him into serious inconsistencies and unsupported assertions, but it also helped him convey important concepts in a remarkably lively way. He became one of the best known and most influential advocates of Maxwellianism in the 1880s, and his work shows more clearly than any other how closely the spread of the new theory was tied, especially at the popular level, to the use of mechanical models.

"We Find Ourselves in a Factory"

In his well-known critique of British physicists' reliance on mechanical models, the French physicist and philosopher Pierre Duhem cited the alarming example of Lodge's *Modern Views of Electricity:*

> Here is a book intended to expound the modern theories of electricity and to expound a new theory. In it there are nothing but strings which move around pulleys, which roll around drums, which go through pearl beads, which carry weights; and tubes which pump water while others swell and contract; toothed wheels which are geared to one another and engage hooks. We thought we were entering the tranquil and neatly ordered abode of reason, but we find ourselves in a factory.[34]

Duhem could not have chosen a better example than *Modern Views* with which to illustrate the British proclivity for devising mechanical models. Lodge's book was the first and one of the most successful popularizations of Maxwellian ideas, and it followed the characteristic British practice of teaching through models. Lodge used illustrative models freely as a way "to fix ideas," to describe new phenomena to his readers in terms of ones with which they were more familiar, and to make his presentation more vivid and engaging. The book was based on a series of articles

---

34. Duhem 1954: 70–71 [1904]. Duhem's book was based on articles published in 1904–5, but the passage quoted here had evidently appeared somewhat earlier; see Thompson 1901: 18–25 for a response to it and a defense of the British model-building tradition.

in *Nature*, which had in turn grown out of a course of popular lectures Lodge had given in the mid-1880s, and though it was greatly refined and expanded along the way, *Modern Views* always retained the flavor of a Victorian scientific lecture, complete with demonstration models and mechanical illustrations.[35]

Duhem's remark that with Lodge one found oneself "in a factory" was more fitting and perhaps less derogatory than he intended. From an early age, Lodge had been fascinated by gears and machinery. "A steam engine," he told John Ruskin in 1885, "is, and especially was, a thing of beauty to me"; as a boy he had loved to watch stationary engines in operation. "I have gloated over a mathematical formula or a piece of machinery, at times," he told Ruskin, "almost as you rejoice in sunshine upon grass"; for Lodge, machines were a source of aesthetic satisfaction.[36] This fascination with machinery was common among late Victorian physicists, who like others of their age and land grew up surrounded by mills and steam locomotives and other products of the Industrial Revolution. William Thomson had an intensely mechanical turn of mind and often treated physics problems in engineering terms, as did Maxwell and many less prominent figures. The ubiquity of machinery in nineteenth-century Britain influenced the basic imagery Victorian physicists chose when expressing their scientific ideas. As Robert Kargon has observed, "The physical content of British theories was drawn not from astronomical analogies as with the French molecularists, but from engineering, from the characteristics of real materials."[37]

Lodge's delight in machinery found expression in the profusion of mechanical models he devised to illustrate various physical principles. His first important paper, published in 1876, was "On a Model for Illustrating Mechanically the Passage of Electricity through Metals, Electrolytes, and Dielectrics, according to Maxwell's Theory"; he repeated and elaborated the account in *Modern Views*.[38] The model consisted of a continuous loop of string running through a series of "buttons" attached by elastic bands to an outer frame (see Figure 4.4). If the buttons were attached firmly to the cord, so that they elastically resisted any displacement, the model represented a dielectric; if they simply rubbed on the cord, allowing it to pass through while generating heat by friction, the model represented a conductor. The properties of the model could be correlated directly with those of a circuit: the coefficient of friction of the

35. Lodge 1889b. The series ran in *Nature* from 6 Oct. 1887 to 31 Jan. 1889; the early installments were based on Dec. 1885 lectures at the London Institution.

36. Lodge to Ruskin, 7 Apr. 1885, OJL-UB; Lodge 1931b: 343; Jolly 1974: 75.

37. Kargon 1969: 430.

38. Lodge 1876b; cf. Lodge 1889b: 32–62, 360–62.

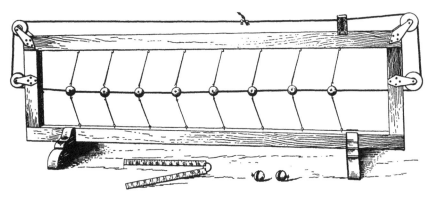

4.4. Lodge's string-and-button model of an electric circuit. The string runs through slots in the buttons, which are attached by rubber bands to the wooden frame. By tightening or loosening the screws holding the buttons to the string, the model can be made to represent either a dielectric or a conducting circuit.

buttons corresponded to the electrical resistivity, the elasticity of the bands to the specific inductive capacity of the dielectric, and so on. Lodge built a working version of his cord model and used it in lectures to illustrate on Maxwellian principles most of the basic phenomena of electricity, particularly those related to dielectric polarization, and to make them clear to an audience for whom a mathematical treatment would have had little meaning.

Lodge also built a "hydraulic model of a Leyden jar" with which to illustrate the charging of a condenser.[39] Borrowing Maxwell's image of an electric current as the flow of an incompressible fluid, Lodge pictured a Leyden jar as an elastic partition (actually a balloon) in a water-filled vessel (see Figure 4.5). By connecting a pump to the vessel, one could distend the balloon, elastically displacing the water and storing energy—just as one could charge a condenser by connecting it to a battery, "displacing" electricity, and storing energy. By opening and closing various stopcocks, one could use the model to imitate the charging, discharging, earthing, and insulating of a condenser and to show how all of these could be explained in Maxwellian terms by polarization of the dielectric without recourse to any action at a distance.

Lodge never actually built a working version of his most famous model, his "cogwheel ether." Unlike his cord model and hydraulic Leyden jar, this model was intended not just to illustrate macroscopic

39. Lodge 1889b: 54–62; cf. p. 315, where Lodge mentions using this model to illustrate a lecture in Dec. 1880.

4.5. Lodge's hydraulic model of a Leyden jar. The glass globe is filled with water and divided by an elastic partition (a balloon). The pressure difference measured by the gauges *a* and *b* corresponds to the voltage difference across a Leyden jar.

electrical phenomena by a mechanical analogue but to represent the ether itself. Lodge derived his model from Maxwell's and FitzGerald's, but instead of using idle wheels or rubber bands to connect his vortices, he pictured an ether built of toothed wheels directly geared together.[40] This gave his diagrams the robust industrial flavor that was to irritate Duhem (see Figure 4.6). Direct gearing meant, however, that adjoining wheels had to spin in opposite directions, whereas to represent a magnetic field, their spins should all be of the same sign. Lodge met this difficulty by dividing the ether into alternating layers of positive and negative wheels and taking a positive wheel spinning in one direction as equivalent to a negative wheel spinning in the other. He thus introduced into his model a fundamental dualism that was not reducible to a purely mechanical difference, a step some of his colleagues found objectionable.

Lodge drew most of the other features of his model quite directly from Maxwell and FitzGerald. He depicted electric displacement by an elastic yielding of the wheels along the direction of the electric force, much as Maxwell had. For conduction, he drew on FitzGerald's model:

40. Lodge 1889b: 178–79, 264.

4.6. Lodge's cogwheel ether, with a moving rack to represent an electric current. The spinning cogwheels represent a magnetic field.

the wheels in a conductor are toothless, Lodge said, so that they partly slip against one another while turning in place and so generate heat by friction, just as FitzGerald's wheels slipped along beneath their bands in a conductor (see Figure 4.7). By combining and modifying Maxwell's and FitzGerald's ideas as needed, Lodge was able to make his model illustrate the main phenomena of Maxwellian electrodynamics in a vivid if not scrupulously consistent way.

Modern Views was widely read, both when it was serialized in *Nature* and when it appeared as a book.[41] Its timing was fortunate. The sections on electromagnetic waves, for example, appeared in *Nature* in mid-1888, just as Hertz and Lodge were demonstrating the existence of such waves experimentally. Interest in the subject ran very high for the next several years, and as the only popularization of Maxwell's theory then available, *Modern Views* benefited. Those who turned to Lodge's book absorbed from it a set of characteristic Maxwellian doctrines: that the ether is the seat of energy and the medium of electrodynamic actions; that the field should be regarded as primary and charges and currents as derivative; and that the ether is ultimately mechanical in nature. Lodge became the leading popularizer of Maxwellian ideas, and his *Modern Views* played a key part in spreading Maxwell's theory beyond the narrow circle of electrical physicists and in making it the property, at least in its main features, of a much wider audience. This popularization was closely tied to the use of models. *Nature* emphasized in its review that the models in *Modern Views* "must prove of the greatest assistance in enabling the student to gain a clear and vivid idea of electrical processes,

41. The first edition was published in London and New York in Sept. 1889, a second in 1892, and a third in 1907. It was also translated into Russian (1889), French (1891), and German (1896). See Besterman 1935, which lists nearly 1,200 of Lodge's publications and is fairly complete.

4.7. Lodge's cogwheel ether, with slippage between toothless wheels to represent dissipative conduction. The toothed wheels represent a layer of dielectric sandwiched between two current-carrying conductors.

and ought to be largely employed by all teachers of electricity."[42] Lodge's readers learned Maxwell's theory through mechanical models and thus also learned to regard such models as the proper route to an understanding of electromagnetism.

The praise for *Modern Views* was not unmixed, and Lodge received considerable criticism for the rather loose way he used his models. As FitzGerald told Lodge in 1893, some of his Dublin colleagues had "objected that they get muddled by your jumping about from one theory to another or rather from one analogy to another instead of presenting some one continuous system that will be consistent throughout."[43] Different models seemed to lead to different conclusions: should an electric current be regarded as a real flow, as in the hydraulic Leyden jar, or simply as a line of slip between spinning bits of the ether, as in the cogwheel model?

FitzGerald subjected the cogwheel ether to especially sharp criticism. "I have been considering your model ether a little this morning," he told Lodge in September 1889, "and the more I think of it the less I like it." His first objection was to Lodge's use of positive and negative wheels:

> It continues the dualistic view of electricity, making another distinction than *mere* difference of displacement between pos[itive] and neg[ative] elect[ricity] while we have no evidence of anything of the kind. . . . That there is no other known difference than one of sign in either elect[ricity] or mag[netism] is a most important part of what I consider a sound modern view of the phenomenon and so I dislike any bolstering up of the old view of two kinds of electricity.[44]

42. *Nature 41* (1889): 5–6.
43. FitzGerald to Lodge, 10 Oct. 1893, OJL-UCL.
44. FitzGerald to Lodge, n.d.; the first sheet (OJL 1/145/3) is in OJL-UB, while the second is in OJL-UCL. Comparison with Lodge to FitzGerald, 1 Oct. 1889, FG-RDS, and other letters shows this one dates from late Sept. 1889.

A similar criticism was voiced in an anonymous review in the *Electrician*.[45] Lodge had simply *defined* positive and negative wheels as opposites, without suggesting any mechanical basis for the difference. His dualistic ether was thus only "quasi-mechanical," the reviewer said, and not in accord with the ideal of mechanical explanation.

FitzGerald also complained to Lodge, "Your model does not on the face of it and explicitly show the connection between electrical and magnetic phenomena." Lodge had depicted electric polarization as a stretching and literal "displacement" of the cogwheels but had not shown how this was connected with the spinning that represented magnetism or the slip that represented conduction current. He had thus obscured one of the central tenets of Maxwell's theory, prompting FitzGerald to exclaim: "Oh! I think your model is horrid!"[46]

In a series of articles in the *Electrician* in 1893, J. H. Poynting extended FitzGerald's critique of Lodge's cogwheel ether and traced the main problems to Lodge's mixing of incompatible models. "It appears to me," Poynting said, "that [Lodge] has grafted on to his model the representation of current in the entirely different model of FitzGerald, and so obtains something quite inconsistent with his previous ideas."[47] Lodge had drawn his representation of dielectric polarization from Maxwell's model, in which the idle-wheel particles were actually displaced by electric forces. He then had to draw his representation of current from Maxwell, too, Poynting said, and admit that a current consisted of a real flow of particles, positive wheels in one direction and negative in the other, rather than trying to introduce FitzGerald's idea that a current was simply a line of slip with no real displacement. To accept Lodge's hybrid would be to accept that there was "a difference in kind between the process of displacing in a dielectric, the equal and opposite motions of the two fluids which are leading to an electrostatic strain, and the current in a conductor." That, Poynting said, would be "rather setting back the clock."[48] The intimate connection between displacement and conduction was the keystone of Maxwell's theory, and both FitzGerald and Poynting took Lodge to task for obscuring it by his mixing of models. Lodge accepted the correction meekly. "I was anxious to show," he said, "that a current *need* be nothing progressive on Maxwell's theory, and perhaps I overstrained this point and made out that it *was* nothing progressive, thereby making my model more difficult to grasp and less sim-

45. *Elec. 23* (1889): 399–400, 423–25; the close parallel with remarks in his letter to Lodge suggests FitzGerald may have written it.
46. FitzGerald to Lodge, [late Sept. 1889], OJL-UB/UCL (see n. 44).
47. Poynting 1920: 250–68 [1893]. These articles grew out of a request that Poynting review *Modern Views*, 2d ed.; see *Elec. 31* (1893): 572.
48. Poynting 1920: 262 [1893].

ply consistent with what had gone before."[49] Lodge's desire to make particular points as vividly as possible had, he admitted, sometimes led him into serious errors.

Of deeper interest than Poynting's particular criticisms of Lodge's cog-wheel ether were his remarks on the proper role of models in physics and the legitimacy of the mechanical program itself. Poynting was more skeptical about mechanical explanations than were most British phys-icists of his day. All such explanations are "merely hypothetical," he said,

> and may at any time have to be discarded. . . . Indeed, they are solely of value as a scaffolding enabling us to build up a permanent structure of facts, i.e., of phenomena affecting our senses. And inasmuch as we may at any time have to replace the old scaffolding by new, more suitable for new parts of the building, it is a mistake to make the scaffolding too solid, and to regard it as permanent and of equal value with the building itself. It is on this point that I find most cause for disagreement with Prof. Lodge. He appears to me to regard his hypothesis as inevitable and permanent, or at least as approximating to the permanent and inevitable. The scaffolding, in fact, is made as important as the building.[50]

Poynting urged great caution in accepting hypothetical models as true reflections of nature. However ingenious the proposed "ethereal ma-chinery" and however closely its behavior fit with known phenomena, there was always the possibility that a new discovery would require it "to be taken off to the lumber-room for worn-out hypotheses" as so many ether models had been in the past.[51] Models could be useful aids to the understanding, Poynting said, but they should not be mistaken for like-nesses of reality.

Poynting viewed the mechanical program itself with considerable skepticism. He was more strongly tinged with positivist and conven-tionalist attitudes toward physical explanation than were most of his British contemporaries. Mechanical explanations were the most satisfy-ing, he said, not because of the nature of the universe but because of the nature of our minds: "It is, I suppose, useless to look for any other than a mechanical hypothesis as final. Probably because we are able to picture mechanical processes, able to think of ourselves as seeing what goes on, seeing kinetic energy manifested in the moving parts, able to think of ourselves as part of the connecting machinery, feeling the stresses, and helping to make the strains, we have come to regard mechanical expla-nations as the inevitable and ultimate ones." But we should not conclude

49. Lodge 1893b.
50. Poynting 1920: 264 [1893].
51. Ibid., p. 255.

that the universe really is essentially mechanical, Poynting said; "for the ether might have properties in its action on matter quite different from any of which we have material types." Poynting viewed models as purely illustrative, not explanatory. He sometimes used illustrative models himself, such as the excellent "turbine-spring" model of the ether he sketched in his *Electrician* articles, but he regarded such models as only very crude analogies, containing not even the shadow of a likeness. "I believe that the time has hardly come for ultimate mechanical construction," Poynting said; for the present, physicists ought to content themselves with formulating electromagnetic laws "and leave the ether out of account."[52]

Lodge was hardly content to "leave the ether out of account" or to defer indefinitely the search for its mechanical structure. Lodge and Poynting stood on opposite sides of a basic conflict about the aims of scientific explanation which went back at least to Descartes and Newton. On one side were those who sought explanation in mechanisms; on the other, those who sought it in laws. Both strands were present in most nineteenth-century British physicists, but Lodge stood far toward the side of mechanisms, Poynting toward the side of laws. The stance a scientist took on this question largely determined his attitude toward the use of models. The prevalence of mechanical models in Victorian physics and the seriousness with which they were regarded indicates that Lodge's was the prevailing position, if in an extreme form.

In his enthusiasm for mechanisms and for teaching by concrete images, Lodge was not always careful to make clear what in his models was to be taken as a true reflection of nature and what was only imaginary. To that extent, Poynting's remark that Lodge confused the scaffolding with the building was justified. But Lodge was under no illusions that his cogwheel model was a true likeness of the ether. When FitzGerald criticized his model in 1889, Lodge replied, "I am sorry you dislike my models so much but am not altogether surprised. I don't greatly hanker after them myself except as a step and better than nothing."[53] Lodge's claim that he did not "greatly hanker after" models was perhaps a bit disingenuous, but it was true that he regarded them as only "a step." As pedagogical tools, his models were a step toward a clearer and more vivid understanding of electromagnetism; as research tools, they offered a step toward a true mechanical likeness of the ether. Lodge knew that cogwheels and factory machinery could be only caricatures of the true mechanism of the ether, but he believed that they could be useful aids in the search for that true mechanism.

52. Ibid., p. 264.
53. Lodge to FitzGerald, 1 Oct. 1889, FG-RDS.

The Vortex Sponge

"What is ether?" That, Lodge declared in 1889, was *"the* question of the physical world at the present time." Believing as he did that the ether underlay all physical phenomena, he regarded the question of its constitution as the most fundamental that could be posed by science. But it was not unanswerable; indeed, Lodge thought in 1889 that it was "not far from being answered." He was quite explicit about what sort of answer he expected and would find acceptable:

> If a continuous incompressible perfect fluid filling all space can be imagined in such a state of motion that it will do all that ether is known to do; if, simply by reason of its state of motion, it can be proved capable of conveying light and of manifesting all electric and magnetic phenomena which do not depend on the presence of matter; and if the state of motion so imagined can be proved stable and such as can readily exist, the theory of free ether is complete.[54]

Lodge's ambition was to find a purely mechanical theory of the ether; the solution he described was the vortex sponge FitzGerald had devised in 1885. This theory was intended to be truly fundamental; it was offered not as a crude mechanical illustration but as a true likeness of the ether. It was perhaps the highest expression of the mechanical program, and from the 1880s through the turn of the century, British physicists expressed considerable optimism that the vortex sponge could indeed be made to provide a purely mechanical theory of the ether and so of the electromagnetic field.

FitzGerald first proposed his vortex sponge ether in a postscript to the same January 1885 letter in which he told Lodge about his wheel-and-band model. "I am just elaborating a theory of the ether," he said, "that requires it to be a perfect liquid chock full of vortices. I can't see any other way of making a liquid possess properties like the ether. It works a long way without a hitch. The principal difficulty is an *embarras de richesses* in the way of hypotheses as to what polarisation of the medium consists in but that can hardly be any objection to the hypothesis which in itself is simplicity itself."[55] That same day, FitzGerald sent a longer account of the sponge theory to J. J. Thomson, who had just bested him in the competition to succeed Lord Rayleigh as Cavendish Professor at Cambridge. Thomson was then known chiefly for his mastery of the intricate mathematical theory of vortices, and FitzGerald asked him "to

54. Lodge 1889b: xi.
55. FitzGerald to Lodge, 1 Jan. 1885, OJL-UCL.

criticise the following scheme of a theory of the ether before I publish lucubrations on the subject."[56]

Vortex mechanics was a lively topic in British mathematical physics toward the end of the nineteenth century, especially at Cambridge. Interest in the subject grew mainly out of William Thomson's suggestion in the 1860s that atoms might be stable vortex rings in a perfect liquid ether, making matter itself simply a "mode of motion" of the all-pervading ether.[57] Later work by Thomson himself and by such younger physicists as J. J. Thomson and W. M. Hicks eventually showed that atoms could not be simple vortex rings, but the study of vortex motion stimulated the idea that the ether might itself be an enormous tangled "sponge" of vortex filaments with matter as closed vortex rings within it.

The vortex theories exemplified the purest and most radical form of the mechanical program, which was especially strong in late Victorian Britain. This "radical mechanist" school put a very strict construction on the word *mechanical:* only pure matter and motion were to be allowed. There could be no recourse to action at a distance or even to elasticity; potential energy itself was to be reduced to the kinetic energy—that is, the motion—of unseen masses.[58] For the radical mechanists, only such a purely mechanical theory could qualify as a true and complete explanation.

FitzGerald made his own radical mechanist leanings clear in his letter to J. J. Thomson. "In the first place I have never been satisfied with Stokes' jelly ether," he said, largely because of its inability to account for electromagnetic phenomena. Moreover, he said, the elastic solid ether left unanswered a fundamental question: "Whence comes its rigidity? It requires internal forces to explain and a complete theory should explain it."[59] As FitzGerald put it elsewhere, "The hypothesis that the ether is like a thin jelly in no way *explains* this property, as it is the possession of properties analogous to rigidity that requires explanation."[60] In a *complete* theory one could not simply posit an elastic ether but would have to explain the apparent elasticity itself in terms of underlying motions. "What physicists ought to look for," FitzGerald declared, "is such a mode of motion in space as will confer upon it the properties required in order

56. FitzGerald to J. J. Thomson, 1 Jan. 1885, JJT-ULC. On the Cavendish professorship, see Rayleigh 1943: 21–22 and FitzGerald to Stokes, 4 Dec. 1884, GGS-ULC.

57. See Smith and Wise 1989: 417–44. William Thomson had suggested a spongelike structure for matter as early as the 1860s and used the term *vortex-sponge* in print in the early 1880s, but FitzGerald and W. M. Hicks were apparently the first to apply the idea to the ether.

58. See Topper 1971 and Klein 1972: 73–75.

59. FitzGerald to J. J. Thomson, 1 Jan. 1885, JJT-ULC.

60. FitzGerald 1902: 154 [1885].

that it may exhibit electromagnetic phenomena. Such a mode of motion would be a real explanation of these phenomena"; anything less would be no more than a "description."[61] The radical mechanists regarded theories that gave only the laws of phenomena, such as Maxwell's of electromagnetism and Newton's of gravitation, as useful but incomplete. For such men as William Thomson, Lodge, and FitzGerald, there was a more fundamental level of reality beneath the phenomenal laws, and they would accept no theory as "complete," no explanation as "real," unless it could trace those laws back to purely mechanical motions in the ether. It was an extremely ambitious program, but as Maxwell said of the vortex atom theory, though "the difficulties of this method are enormous . . . the glory of surmounting them would be unique."[62]

FitzGerald's support for radical mechanism was reinforced by his adherence to a rather idiosyncratic version of Berkeley's idealist philosophy. It is hard to say how important this philosophy was in forming FitzGerald's mechanist views, especially since others who arrived at essentially the same position made no reference to Berkeley's philosophy or even were hostile to it. But it is clear from remarks in FitzGerald's writings that he took his interpretation of Berkeley quite seriously and believed that it provided fundamental metaphysical support for his contention that the physical world was reducible to pure motion.

Berkeley had been a Trinity College Dublin man, and his ideas retained some currency there throughout the eighteenth and nineteenth centuries. Metaphysics was among the subjects FitzGerald studied for his fellowship examinations, and it was reportedly then that he was first attracted to Berkeley's philosophy. According to his brother Maurice, FitzGerald's "metaphysical bent came from both sides of the family"; his father had taught moral philosophy at Trinity, and "both he and Johnstone Stoney and that family [were] given to metaphysics."[63] Stoney wrote about his own philosophical views at some length, and FitzGerald apparently derived his metaphysical argument for the kinetic nature of phenomena at least in part from his uncle.

FitzGerald's own published remarks on Berkeley and metaphysics are scattered and brief. The longest came toward the end of his 1896 Helmholtz Lecture to the Chemical Society of London. After discussing Helmholtz's work on vortex mechanics, FitzGerald declared:

61. Ibid., p. 162 [1885].
62. Maxwell 1890, 2:472 [1875]. William Thomson and Maxwell had radical mechanist leanings but were not as thorough in following them out as were some of their successors.
63. Maurice FitzGerald to Larmor, 4 Mar. 1901, JL-RS; cf. Lodge 1902, in FitzGerald 1902: xxxiii.

We cannot help being impressed with how far ultimate explanations of nature lead us closer and closer to the conclusion that these phenomena of our consciousness are all explicable as differences of motion. It is the motion which is imposed upon us. Is there not, then, reason in the suggestion that colour and sound, nay, space, time, and substance are functions of our consciousness, produced by it under the action of what may be called an external stimulus, and that the only part of the phenomenon which essentially corresponds to that stimulus is the always pervading motion? And what is the inner aspect of motion? In the only place where we can hope to answer this question, in our brains, thought is the internal aspect of motion. Is it not reasonable to hold, with the great and good Bishop Berkeley, that thought underlies all motion?[64]

FitzGerald put the same point more briefly in an 1890 lecture at the Royal Institution and made explicit its connection to the vortex sponge theory:

This hypothesis explains the differences in Nature as differences of motion. If it be true, ether, matter, gold, air, wood, brains, are but different motions. Where alone we can know what motion in itself is—that is, in our own brains—we *know* nothing but thought. Can we resist the conclusion that all motion is thought? Not that contradiction in terms, unconscious thought, but living thought; that all Nature is the language of One in whom we live, and move, and have our being.[65]

That is, when we examine nature, we seem to find that everything can be resolved into motion; even our brains, if examined as objects, would presumably appear as nothing but very complex motions. But we also subjectively experience this motion in our brains as thought, suggesting—irresistibly, in FitzGerald's view—that all motion is the objective manifestation of thought: God's thought, of course. Our sensations are thus caused not by an external world of self-existent matter but by God's thoughts; apparent material objects are simply steps in the process by which we apprehend God's thoughts. There is thus no dualism between mind and matter, and thought appears not as a mysterious concomitant of certain motions in our brains but as the one true reality from which all appearances are derived.

Stoney stated this argument in some detail in 1885 and again in 1890.[66] For him, as for Berkeley, regarding the phenomenal world as a

---

64. FitzGerald 1902: 376 [1896].
65. Ibid., p. 276 [1890].
66. Stoney 1885, 1890b.

projection of the thoughts of God was a way to guard against materialism and atheism. Ironically, however, Stoney's emphasis on *motion* as the physical manifestation of God's thought led him to a position that for practical purposes was quite close to the materialists' own. Whether our sensations are caused by the actual motion of material bodies or by our apprehension of the thoughts of God was irrelevant for most scientific purposes. Our observations would be the same. But Stoney's approach did have one important scientific implication: since he denied the existence of material substance, he had to regard the primary phenomenon of the physical world as the motion of portions of space itself. The "elemental ether," Stoney said, was "space itself regarded as movable."[67] In practice, this movable space was equivalent to an all-pervading fluid ether of uniform density. Stoney proposed to reconstruct the entire physical world purely from the motions of this elemental ether. His idealist philosophy had led him to a radical mechanist position.

Stoney gave his clearest public statement of this doctrine of motion as the basis of the physical world in February 1885, just as FitzGerald was working out his vortex sponge theory. There can be little doubt that the two men had discussed Stoney's metaphysical argument before 1885, and there is every indication that FitzGerald was convinced by it. Certainly his vortex sponge ether was entirely consistent with Stoney's strictures, and it is worth noting that in his 28 March 1885 paper at the Physical Society in London, FitzGerald had called upon physicists to look for "such a mode of motion in space as will confer upon it" (i.e., upon *space*) electromagnetic properties, thus echoing the notion of "movable space" that Stoney had proposed at the Royal Institution just a few weeks before.

FitzGerald's main argument for the vortex sponge was very general: it was, he said, the only purely mechanical theory that could be based on a continuous ether. Maxwell's electromagnetic ether had to be rigid enough to sustain strong shearing stresses. No ordinary motionless liquid could do this, and as FitzGerald observed to J. J. Thomson, "It is evident that unless we can explain the apparent rigidity of the ether by supposing it either in irrotational or rotational motion it is all up with a liquid ether."[68] The only alternative to a vortex sponge was some type of elastic solid theory in which rigidity was simply assumed, and to FitzGerald that was no explanation at all. He had strong reasons, both physical and metaphysical, for taking a liquid plenum as the only acceptable basis for a truly ultimate theory of the ether, and he was convinced

67. Stoney 1890a: 403.
68. FitzGerald to J. J. Thomson, 1 Jan. 1885, JJT-ULC.

that "the only way in which to impart to a perfect liquid properties at all analogous to those possessed by the ether, is by supposing it full of motion."[69] William Thomson had shown that such motion could make a mass of liquid resist deformation and otherwise act much like a solid body, and as FitzGerald remarked, "if vortices can make a small piece of a strong elastic solid, we can make watches and build steam engines and any amount of complex machinery," so that there seemed to be no limit to what a vortex ether might hope to explain. There were, FitzGerald said, "almost infinite possibilities in a vortex sponge."[70]

The evident explanatory power of the vortex ether also attracted others, notably W. M. Hicks, who proposed a vortex sponge somewhat similar to FitzGerald's in September 1885. At the 1887 British Association meeting in Manchester, William Thomson took the work of FitzGerald and Hicks further in a brief paper "On the Vortex Theory of the Luminiferous Aether."[71] In this "mathematical exercitation," as Lodge called it, Thomson showed that a vortex sponge would be able to convey transverse waves like those of light. Both Hicks and FitzGerald were present and pronounced themselves satisfied with the proof and "conscious of the greatness of the discovery." Indeed, it was the apparent fulfillment of a large part of the radical mechanist program; FitzGerald reportedly called it "the greatest step towards the comprehension of the intrinsic structure of the universe which had been made since the time of Newton." Lodge, by then a full convert to the sponge theory, hailed Thomson's paper as "the thing for which in all future times the visit of the Association to Manchester will probably be famous."[72] FitzGerald and Lodge were perhaps extravagant in their praise, but their remarks indicate the enthusiasm the vortex sponge theory could arouse among those who were seeking a truly ultimate mechanical theory of the universe.

This enthusiasm rose even higher in May 1889 when FitzGerald sent *Nature* a letter "On an Electromagnetic Interpretation of Turbulent Liquid Motion." In this highly mathematical piece, evidently dashed off in a burst of inspiration, FitzGerald claimed to show that the whirls and flows of a vortex sponge could be consistently correlated with the electric and magnetic vectors of Maxwell's theory, thus completing the identification of the electromagnetic and vortex sponge ethers that he had first suggested four years before.[73] Lodge, who was then writing the preface to the first edition of his *Modern Views*, inserted an enthusiastic reference to

69. FitzGerald 1902: 156 [1885].
70. Ibid., p. 239 [1888]; FitzGerald to J. J. Thomson, 1 Jan. 1885, JJT-ULC.
71. William Thomson 1887; Hicks 1885.
72. Lodge 1887: 313.
73. FitzGerald 1902: 255–61 [1889].

FitzGerald's letter. By devising a vortex ether "which is not only optically, but also electrically, sufficient," FitzGerald had made an important advance on William Thomson's 1887 theory, Lodge declared. "If no flaw appears," he said, "if it stand the test of criticism and further development, the theory of free ether is far more than begun."[74] The spring of 1889 was an exciting time for FitzGerald and Lodge. Their radical mechanist program had, they believed, won a great victory; "the intrinsic structure of the universe" was at last, it seemed, being laid bare.

This optimism was soon tempered. Errors and oversights were found in FitzGerald's 1889 paper, and it soon became clear that the whirls and flows of the ether could not be identified with the electromagnetic vectors as directly as he had thought and, further, that the relatively simple sponge he had proposed was unstable and would soon lose its ability to convey waves or exert electromagnetic forces. William Thomson pointed out most of these problems in letters to FitzGerald in 1889 and 1890. At first optimistic about the scheme, Thomson found in September 1889 that FitzGerald's sponge of ordinary vortices would be unstable and that only vortices with vacuous cores could persist.[75] FitzGerald did not find such hollow vortices very appealing, perhaps because they conflicted with Stoney's doctrine of a plenum of "movable space." In any case, as 1890 wore on and Thomson refuted, one by one, all of his attempts to defend the original theory, FitzGerald grew discouraged about the immediate prospects for the vortex sponge. "I heard from Sir Wm. T.," he told Lodge in June 1890, "and my vortex solution won't work. I hardly thought it would but I overlooked a rather obvious objection which is not encouraging."[76]

FitzGerald tried to salvage his theory at the 1890 British Association meeting by showing that with very thin vortices the tendency to instability could be indefinitely delayed and that the breakup of the sponge would in fact tend to become slower and slower as the vortices spun around each other and pulled themselves thinner. But though Thomson agreed that this would occur with two vortex filaments, he declared "in an oracular manner that it would not be generally true. I collapsed," FitzGerald told Heaviside, "and have not ventured to publish and am afraid to ask him about it."[77] The combination of discourage-

---

74. Lodge 1889b: xii [dated 13 May 1889]; retained unchanged in Lodge 1892b.
75. William Thomson to FitzGerald, 28 Sept. 1889, 29 Mar. 1890, and 12 June 1890, FG-RDS; see also Larmor's footnotes to FitzGerald 1902: 255–61.
76. FitzGerald to Lodge, 19 June 1890, OJL-UCL.
77. FitzGerald to Heaviside, 25 Aug. 1893, OH-IEE. Only the title of FitzGerald's "Note on the Relation between the Diffusion of Motion and Propagation of Disturbance in Some Turbulent Liquid Motions" appeared in BA Report (1890), p. 757.

ment and hope for the future that the proponents of the vortex sponge felt in 1890 comes across very clearly in Lodge's remarks on FitzGerald's British Association paper:

> At present the vortex theory of the ether labours under a great many difficulties, for . . . it threatens to gradually change its properties with time, containing within itself the seeds of its own entanglement. Problems like this, however, are not to be solved in a few years; and, though present prospects of the theory look dark, some further discovery may suddenly change the aspect of affairs, and verify what at present can only be regarded as a brilliant speculation.[78]

FitzGerald expressed his own feelings about both the difficulties and the promise of the vortex sponge in a revealing way in an 1893 letter to Heaviside, who was then only vaguely familiar with the theory. It was a pity, FitzGerald said, "that someone with a store of patience and energy for indefinite calculation does not work on the average motion of liquids. There is evidently a field quite comparable to that of the average motion of particles which is the kinetic theory of gases and I feel horribly confident that it is *the* theory of the ether. I have a sort of feeling in my bones that it must be so: I suppose I am a bit a crank upon it."[79] It was this "feeling in his bones" that underlay FitzGerald's continued advocacy of the vortex sponge, a feeling akin to that Lodge once expressed about Thomson's vortex atom theory. "Is it not highly beautiful?" he had said in an 1882 lecture. "A theory about which one may almost dare to say that it deserves to be true?"[80] The vortex atom and its offspring, the vortex sponge, were such grandly beautiful theories that even when beset with grave difficulties, they continued to attract the allegiance of some of Britain's best physicists.

The vortex sponge was a peculiarly seductive hypothesis: it offered natural philosophers a theory of the universe that was complete, unified, and yet fundamentally very simple. There was, however, so much room for complication in working out its consequences that the sponge seemed capable of accommodating virtually anything, and vortex mechanics was so difficult and intricate a subject that no one could say with assurance that any proposed vortex explanation was impossible. To one committed to the mechanical program and so to an extremely strict standard of "ultimate" explanation, the vortex sponge thus held an attraction that was not easily displaced. As late as 1931, Lodge looked back on

78. Lodge 1890a: 575.
79. FitzGerald to Heaviside, 25 Aug. 1893, OH-IEE.
80. Lodge 1889b: 356 [1882].

the vortex theories of the 1880s and 1890s and said that he still expected "posterity will recognize some inklings of truth in these hydrodynamical speculations."[81] FitzGerald, had he lived, would almost certainly have agreed.

"Mathematical Machinery"

The goal of understanding electromagnetism mechanically was eventually abandoned by most physicists, and the search for such "likenesses" of the ether as the vortex sponge gradually faded as Maxwell's equations of the electromagnetic field came to be regarded as themselves fundamental and self-sufficient. Einstein described the process many years later: "One got used to operating with these fields as independent substances," he said, "without finding it necessary to give one's self an account of their mechanical nature; thus mechanics as the basis of physics was being abandoned, almost unnoticeably, because its adaptability to the facts presented itself finally as hopeless."[82] Physicists ceased to feel a need to look for a mechanism behind the electromagnetic laws or to believe that their understanding would be improved by finding one.

This attitude did not prevail in Britain until well into the twentieth century, but there were earlier signs, especially in the work of Heaviside and Poynting. Heaviside affirmed many times that there must be a mechanical ether; to deny it, he said, would be "thoroughly anti-Newtonian, anti-Faradaic, and anti-Maxwellian."[83] But attempting to find the actual structure of the ether was, he thought, too ambitious and speculative a task to be worthwhile. "What a difficult piece of work it would be," he told FitzGerald, to show that Maxwell's laws could be derived from the motions of a perfect liquid. "It may be so," he said, "but *I* certainly don't see my way to it at present."[84]

In his own work, Heaviside was usually content simply to use Maxwell's equations without inquiring into a possible underlying mechanism. Devising models did not attract him. "I am myself shy of models for two reasons," he told FitzGerald. "First because I am not mechanical to any important extent, and next because as regards the working out of electrical and especially electromagnetic problems it is easier to follow the behaviour of tubes of displacement and induction than the corre-

81. Lodge 1931a: 85.
82. Einstein, in Schilpp 1949, 1:27.
83. Heaviside 1893–1912, 3:479 [1911].
84. Heaviside to FitzGerald, 30 Aug. 1893, FG-RDS.

sponding quantities in a model." As he put it more briefly in a letter to Hertz, "My experience of so-called 'models' is that they are harder to understand than the equations of motion!"[85]

The process of learning to be content with purely electromagnetic explanations was eased by the evolution of Maxwell's equations into a "mathematical model" of the electromagnetic field. Heaviside in particular deliberately tailored what he called his "mathematical machinery" to reflect as closely as possible the physical nature of electromagnetic phenomena.[86] He developed his system of vector analysis, which has since proven of enormous value in nearly all areas of physics, mainly as a way to give a clear and vivid representation of the electromagnetic field and of the behavior of its "tubes of displacement and induction."

Heaviside derived his vector system from the quaternions William Rowan Hamilton invented in 1843. A quaternion consists of an ordinary number, or "scalar," combined with a "vector," which has both magnitude and direction. The vector can in turn be broken down into three perpendicular components, the total of four elements giving the quaternion its name. The unusual algebraic properties of quaternions were Hamilton's main interest, but the correspondence between their vector parts and the three dimensions of space made them well suited to treating physical problems, as P. G. Tait showed in his 1867 *Elementary Treatise on Quaternions*. Quaternions were taken up by a few other physicists, mostly in Ireland (FitzGerald was, he said, "'riz' on Tait"), and Tait's friend Maxwell found the idea of directed quantities useful enough that he gave the main results in his *Treatise* in quaternionic as well as Cartesian form.[87]

When Heaviside read Maxwell's *Treatise* in the 1870s, he was impressed by the power and compactness of the quaternion notation and turned to Tait's *Elementary Treatise* for more information. He found it, however, "without exception the hardest book to read I ever saw."[88] Even after mastering the subject, he continued to find the rules of quaternion algebra troublesome, particularly the one requiring the square of a vector to be negative. Moreover, he said, "on proceeding to apply quaternionics to the development of electrical theory, I found it very inconvenient. Quaternionics was in its vectorial aspects antiphysical and unnatural and did not harmonise with common scalar mathematics. So I dropped out the quaternion altogether, and kept to pure scalars and

85. Ibid., 28 June; Heaviside to Hertz, 14 Aug. 1889, HH-DM.
86. Heaviside 1893–1912, 1:214 [1892].
87. Tait 1867; FitzGerald to Heaviside, 26 Sept. 1892, OH-IEE. For an excellent account of the development of vector analysis, see Crowe 1967.
88. Heaviside to Lodge, 13 Apr. 1891, OJL-UCL.

vectors, using a very simple vectorial algebra in my papers from 1883 onward."[89] Vectors and scalars and their products are very common in physics, but full quaternions and quaternion products are rare. By breaking the quaternion into its scalar and vector parts and giving straightforward rules for the addition and multiplication of vectors, Heaviside was able to free vectors from their quaternionic baggage and produce a simple and flexible vectorial system that could be readily applied to physical problems—most notably, of course, in Maxwell's electromagnetic theory. Together with J. W. Gibbs in America, who had independently followed a very similar path, Heaviside took the lead in promoting vectors in the late 1880s and 1890s. The clarity and utility of his vector system, especially in electromagnetic problems, soon enabled it to prevail over both the Cartesian and quaternionic systems, and vector analysis is now taught to all beginning physics students. The notation usually used today is Gibbs's, but the fundamental ideas are the same as Heaviside's.[90]

With his vector tools in hand, Heaviside turned to the electromagnetic problems for which he had fashioned them in the first place. Beginning in the *Electrician* in 1882–83, he produced a long series of papers explaining how electromagnetic phenomena could be treated in terms of "vector fields," in particular by using the fundamental theorems of "slope" (now called "gradient"), "divergence," and "curl." He followed Maxwell's basic physical ideas quite closely but gave them a clearer and simpler form. He was smoothing the way, he said, for those who would follow: "To use a simile which may be readily applied, when once a bold explorer has reached the top of a hitherto inaccessible mountain, by circuitous paths and with the expenditure of great labour, others can do it after him, and easier routes may be discovered; and in course of time people may travel up comfortably seated in railway carriages."[91] By defining vector quantities and operations in particular ways, Heaviside was able, as W. E. Sumpner observed, "to develop a mathematical machine specially suitable" for use with Maxwell's theory.[92] A mathematical investigation expressed in the new vector symbols was not simply a string of syllogisms leading from premises to conclusions; it became, as Pierre Duhem acutely remarked, a model by which the actions of the field could be accurately represented at each step.[93] One could almost see, depicted in the equations, the tubes of displacement diverging from an

89. Heaviside 1893–1912, 3:136 [1902].
90. Crowe 1967: 162–220, 242.
91. Heaviside 1892, 1:214 [1882].
92. Sumpner 1932: 849.
93. Duhem 1954: 76–80 [1904]; cf. Thompson 1901: 24.

electric charge, or the lines of magnetic force curling around a current-carrying wire. With Heaviside's vector methods, one could portray the electric and magnetic fields and their interactions in a direct and almost palpable way; it was no longer necessary to invoke a mechanical model to give the electromagnetic equations physical content. Mechanical models like Lodge's cogwheels or FitzGerald's wheels and bands became dispensable, and in time the search for such "likenesses" as the vortex sponge also faded, as the new mathematical machinery of Maxwell's theory came to seem as real and intelligible as any unseen mechanisms in the ether.

CHAPTER FIVE

"Maxwell Redressed"

Maxwell's was an immensely fertile theory, but in the form in which it appeared in his *Treatise* it was also awkward, confusing, and on some points simply wrong. It was especially ill-suited to handling the propagation problems that were coming into increasing prominence in the 1880s in connection with advances in telegraphy, telephony, and the study of electromagnetic waves. Before Maxwell's theory could gain wide acceptance and come into general use, it required substantial revision and clarification; both its physical principles and their mathematical expression had to be put into a simpler and more easily grasped form. The most important steps in this process were taken in the mid-1880s in the wake of work by Poynting, FitzGerald, and Heaviside on the flow of energy in the electromagnetic field. As the Maxwellians came to focus more and more closely on the physical state of the field at each point in space, they sought to banish the vestiges of the old action-at-a-distance approach that were embodied in Maxwell's continued use of the vector and scalar potentials. Heaviside in particular came to regard the electric and magnetic forces and fluxes as far more fundamental than the potentials, and in 1884 and 1885 he recast the long list of equations Maxwell had given in his *Treatise* into the compact and symmetrical set of four vector equations now universally known as "Maxwell's equations." Reformulated in this way, Maxwell's theory became a powerful and efficient tool for the treatment of propagation problems, and it was in this new form—"Maxwell Redressed," as Heaviside called it—that the theory eventually passed into general circulation in the 1890s.[1]

1. Undated note on the back of p. 5 of Heaviside's manuscript of "Theory of Voltaic Action," Box 14, OH-IEE.

Energy Paths

The guiding principle in the revision of Maxwell's theory in the 1880s was the concept of energy flow. The idea that energy moves along identifiable paths through an electromagnetic field and that Maxwell's theory could be used to delineate those paths was first formulated by J. H. Poynting late in 1883, then rediscovered independently a few months later by Oliver Heaviside.[2] At a time when work on Maxwell's theory could easily have wandered off into purely mathematical elaborations, the discovery of the energy flux theorem focused attention firmly on the physical state of the field. FitzGerald drew attention to the strategic importance of the theorem in an 1892 letter to Heaviside: "One of the points I always look at in the El. Mag. Theories," he said, "is that one of Poynting's—it seems to me to distinguish the Physicist from the Old Tory who sticks to the old ideas on the one hand and the Mathematician . . . on the other."[3] The flux theorem provided a new axis around which Maxwell's theory could be rebuilt, and attention to the flow of energy became one of the hallmarks of the emerging Maxwellian synthesis.

That the electromagnetic field could store and convey energy was scarcely a new idea in 1883; it had been an important part of field theory since the time of Faraday and had been explored in considerable detail by William Thomson in the late 1840s.[4] The phenomena of induced currents observable in any telegraph office or physics laboratory made it clear that an electric current possessed a certain amount of energy and also demonstrated that this energy could pass across seemingly empty space, as in an air-core transformer. The behavior of condensers made much the same point about electrostatic energy. When Maxwell set about constructing his new electromagnetic theory in the 1860s, he made the energy associated with charges and currents one of its foundations.

It was not at all clear just where this energy was located, however; for its distribution could be expressed in alternative ways that embodied radically different views of the nature of electrodynamic actions. On the one hand, the energy of a current could be expressed as the integral over space of $\frac{1}{2}\mathbf{A} \cdot \mathbf{C}$, the product of the vector potential and the current densi-

2. Poynting 1920: 175–93 [1884]. This paper was received by the Royal Society on 17 Dec. 1883 but not published until after 19 June 1884, the date of a final footnote (p. 184n). An abstract consisting of its first four paragraphs was published in *Proc. RS 36* (10 Jan. 1884): 186–87, undercutting the suggestion in Nahin 1988: 118 that Heaviside's June 1884 statement of the flux formula was the first actually to appear in print.

3. FitzGerald to Heaviside, 26 Sept. 1892, OH-IEE.

4. See Wise 1979.

ty. This formula located all the energy of a steady current within the wire, since there was no current outside it. It thus took no account of any intervening medium and was, as Maxwell pointed out, "the natural expression of the theory which supposes the currents to act upon each other directly at a distance."[5] But the energy of the current could also be expressed as $\frac{1}{2}\mu H^2$, where μ is the permeability and H the magnetic force. When integrated over all space, this expression gives the same value for the total energy of the current as the $\frac{1}{2}A \cdot C$ form, simply in consequence of the defined mathematical relations among A, C, and H. It locates the energy very differently, however: according to the $\frac{1}{2}\mu H^2$ formula, almost all the energy is *outside* the wire, in the surrounding magnetic field. This expression shifts attention away from the current itself and toward its associated field; thus, Maxwell said, it was the form "appropriate to the theory which endeavours to explain the action between currents by means of some intermediate action in the space between them"—that is, the field theory Faraday originated and Maxwell himself elaborated.[6] Similar arguments applied to the choice between expressing the electrostatic energy in terms of the charge and the scalar potential, as $\frac{1}{2}q\psi$, or in terms of the permittivity and the electric force, as $\frac{1}{2}\epsilon E^2$.

Although Maxwell professed to regard the $\frac{1}{2}\mu H^2$ and $\frac{1}{2}\epsilon E^2$ formulas as fundamental, he in fact made relatively little use of them in his *Treatise* and did not pursue very thoroughly the implications of the energy distribution they indicated.[7] Instead, he generally treated the potentials A and ψ as fundamental quantities and thus retained many aspects of the action-at-a-distance approach. He later explained that in an effort to soften the novelty of his *Treatise*, he had "sometimes made use of methods which I do not think the best in themselves, but without which the student cannot follow the investigations of the founders of the Mathematical Theory of Electricity."[8] By continuing to use the potentials, Maxwell obscured some of the most important consequences of his own theory, particularly those concerned with the localization and propagation of energy in the field.

    5. Maxwell 1873, art. 636.
    6. Ibid., art. 606; cf. art 636.
    7. The relationship between the distribution of energy and the stress in the field is discussed briefly in Maxwell 1873, arts. 110–11 and 641–42; cf. Buchwald 1985a: 34–37, 43.
    8. From Maxwell's posthumous *Elementary Treatise on Electricity* (1881), quoted by W. D. Niven in his introduction to Maxwell 1890, 1:xxix. Maxwell added that he had later "become more convinced of the superiority of methods akin to those of Faraday" and had used them from the first in his *Elementary Treatise*, but he never put them into full mathematical form.

The distribution and flow of electromagnetic energy attracted relatively little attention among Maxwell's initial group of followers at Cambridge. Although he was a prominent professor and the first director of the Cavendish Laboratory, Maxwell did not attempt to found a "research school" devoted to exploring and promoting his own theories. Instead he encouraged his students to follow their own inclinations, and many pursued topics far removed from Maxwell's own work.[9] Even those of his students and colleagues who did study Maxwell's *Treatise* in the 1870s generally sought to assimilate it within the Cambridge Tripos tradition and were more intent on mining it for abstruse mathematical problems than on pursuing its underlying physical meaning.[10]

John Henry Poynting (1852–1914) was in some ways an exception to this pattern. The son of a Unitarian minister and schoolmaster, he attended Owens College in Manchester before going up to Cambridge in 1872. He did well on the Tripos examination, finishing as third wrangler in 1876, but leaned more toward physics than mathematics; as Joseph Larmor later observed, Poynting's "interest seems always to have lain in the direct elucidation of physical laws and principles rather than in the evolution and exposition of their consequences by analysis."[11] After a stint as a physics instructor back at Owens College, he accepted a fellowship at Trinity College in 1878 and returned to Cambridge. There he became one of the few important Maxwellians to work directly under Maxwell—though his research at the Cavendish concerned the measurement of the gravitational constant and had nothing to do with electromagnetism.[12]

In 1880, Poynting was appointed professor of physics at the new Mason College of Science in Birmingham, where he soon embarked on a closer study of Maxwell's theory, in particular the question of how energy is distributed in an electromagnetic field. He realized that if one took the conservation of energy seriously and applied this law not just globally but to each local region of space at each moment in time, then the law implied the *continuity* of energy and so the possibility of tracing its passage from place to place. Given that the electromagnetic field can store and convey energy, Poynting said, "we are naturally led to consider

9. Schuster 1910: 38.

10. On the early reaction of Cambridge mathematicians to Maxwell's *Treatise*, see the remarks in Whittaker 1951: 310 on the work of Charles Niven and Horace Lamb; cf. W. D. Niven's comment in his introduction to Maxwell 1890, 1:xxix, that the *Treatise* was regarded at Cambridge as "remarkable for the handling of the mathematical details no less than for the exposition of physical principles."

11. Larmor 1914, in Poynting 1920: xxiv.

12. Schuster 1910: 30. Poynting was at the Cavendish for only a short time before Maxwell died.

the problem: How does the energy about an electric current pass from point to point—that is, by what paths and according to what law does it travel from the part of the circuit where it is first recognizable as electric and magnetic to the parts where it is changed into heat and other forms?"[13]

Poynting proceeded to attack the problem in a very straightforward way, starting with Maxwell's expressions of $\frac{1}{2} \epsilon E^2$ and $\frac{1}{2} \mu H^2$ for the densities of electrical and magnetic energy and calculating how these would change within a specified volume. This was not as easy as it might sound, however, and Poynting had to go through a long series of integrations and transformations, including a detour through the vector and scalar potentials, before he was able to reduce his expression to a few distinct and meaningful terms. The final result, presented to the Royal Society in December 1883 and published in the *Philosophical Transactions* in the latter half of 1884, was surprisingly simple: the rate at which energy enters a region, there to be stored in the field or dissipated as heat, depends solely, Poynting said, on the values of the electric and magnetic forces at its boundaries. The energy flux at each point in space—the "Poynting vector," as it has come to be called—is simply the vector product of the electric and magnetic forces at that point: $S = E \times H$.[14] It was a remarkable discovery, and one that soon put much of electromagnetic theory in a new light.

When Poynting proceeded to apply his flux theorem to a few simple cases, he found some surprising results. Consider a straight wire carrying a steady current. Within and immediately outside the wire, the lines of electric force E are parallel to the current, while the lines of magnetic force H form circles centered on the axis of the wire, their direction determined by the usual "right-hand rule."[15] The Poynting vector $S = E \times H$ thus points radially into the wire, indicating that the energy does not flow along *within* the wire, as everyone had always assumed, but instead streams through the surrounding dielectric and enters the wire through its sides (see Figure 5.1). We must abandon the "prevailing and somewhat vague opinion" that energy is "in some way . . . carried along the conductor by the current," Poynting said, and recognize the sur-

13. Poynting 1920: 175–76 [1884].
14. Ibid., pp. 176–77. Poynting did not express the flux in this vector form, though he gave some results in quaternions.
15. As pointed out in Heaviside 1892, 2:94 [1887] (cf. Nahin 1988: 132), the electric field outside an ordinary current-carrying wire is in fact nearly perpendicular to its surface, with only a small component running parallel to the wire. Poynting was referring, however, to the field immediately outside a short wire of very high resistance, a case in which the parallel component predominates.

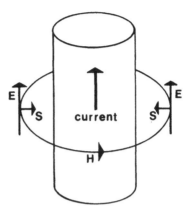

5.1. The Poynting flux of energy
(S = E × H) into a current-carrying
wire.

rounding air or insulation as the true conveyor of electromagnetic energy.[16]

Poynting next used his theorem to analyze the flow of energy in the
neighborhood of a condenser discharged through a high-resistance
wire.[17] In such a discharge, a conduction current flows along the wire
from the positive plate A to the negative plate B, while the subsidence of
the electric field between the plates gives rise to a displacement current
running from B to A (see Figure 5.2). The circuit is thus complete, but
note that the displacement current $\dot{D}$ runs *against* the direction of the
electric force E between the plates, while the conduction current runs
with the direction of the electric force in the wire. Since the direction of
the magnetic force H is determined by that of the current, the Poynting
flux S = E × H points *out* from between the plates and *in* around the
wire. When the discharge begins, the energy that was initially stored as
electrostatic strain within the condenser flows outward across the field
(along the equipotential surfaces, if the discharge is slow enough not to
disturb them) and converges on the wire. The wire acts not as a conduit
for the flow of energy, but simply as a sink within which some of the
strain in the field is broken down and dissipated as heat.

Poynting's flux theorem flew in the face of commonsense notions
about how energy is conveyed by electric currents, and many old-line
electricians rejected it as an absurd and useless hypothesis.[18] But to the
Maxwellians, it seemed simply a natural extension of Faraday's and Max-

16. Poynting 1920: 192 [1884].
17. Ibid., pp. 183–84. The high resistance makes the discharge slow enough to avoid
complicated oscillatory effects.
18. The dismissive remarks of S. A. Varley and John Sprague quoted in Hunt 1983:
350–51 were typical.

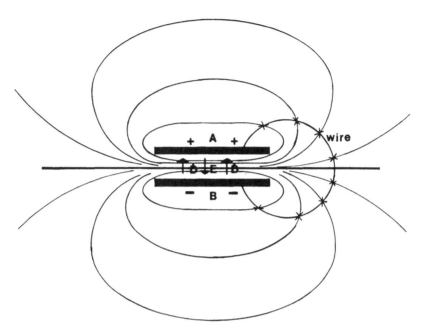

5.2. The Poynting flux of energy from a condenser into a discharging wire. As the displacement subsides, energy flows out from the space between the plates and radially into the discharging wire.

well's ideas on the primacy of the field and its role as a storehouse of energy. Poynting spoke for a generation of British physicists when he declared, "According to Maxwell's theory, currents consist essentially in a certain distribution of energy in and around a conductor, accompanied by the transformation and consequent movement of energy through the field."[19] The Maxwellians already regarded the field, not the current, as the real seat of electromagnetic activity, and by reinforcing and giving substance to this basic tenet of field theory, the flux theorem gave a fresh impetus to their program.

Model Research

FitzGerald was one of the first to recognize the true importance of "Poynting's great discovery," as he called it, and it had a far-reaching effect on the way he thought about electromagnetic propagation.[20] This

19. Poynting 1920: 175 [1884].
20. FitzGerald to Lodge, 1 Jan. 1885, OJL-UCL. Lodge did not at first appreciate the importance of Poynting's discovery; see Lodge to FitzGerald, 3 Jan. 1885, FG-RDS.

influence made itself felt mainly through the wheel-and-band model he devised in 1885. As we saw in the last chapter, FitzGerald designed this model specifically to provide a concrete embodiment of Poynting's flux theorem; it was by analyzing its workings that he first gained a solid grasp of how force and energy propagate in an electromagnetic field, which in turn led him to make a subtle but important revision in Maxwell's basic equations.

The way FitzGerald used his model comes across most clearly in a forty-page section of one of his surviving research notebooks.[21] In these pages, apparently written in the summer of 1888, he attacked propagation questions along two parallel paths, one purely mathematical and one based on the workings of his model. As he sought to trace a detailed correspondence between the equations governing his model and those of Maxwell's theory, FitzGerald filled his notebook with diagrams of wheels and bands and mathematical analyses of how these would work under various circumstances. He then used insights drawn from the model to modify and correct Maxwell's electromagnetic equations. FitzGerald trusted his wheel-and-band model and used it as a genuine research tool, not just to illustrate Maxwell's theory, but to revise and improve it.

The potential functions Maxwell had used in his *Treatise* were ill-suited to treating propagation problems; in particular, his equations implied that the electric or scalar potential ψ was propagated instantaneously.[22] This seemed much like direct action at a distance and led to considerable confusion in the 1870s and 1880s about the nature of propagation in Maxwell's theory. FitzGerald had himself been led astray on this question in his first papers on electromagnetic waves in 1879 and 1880, and though his introduction of a retarded form of the vector potential $\mathbf{A}$ in 1882 had corrected part of the problem, it did nothing to alter the presumption that the electric potential ψ was propagated instantaneously.

The fundamental insight FitzGerald drew from his model was quite simple, and he stated it early on in his notebook. Just three pages into his discussion, he interrupted some calculations on the relation between ψ and $\mathbf{A}$ to remark, "Still there remains that the wheels seem to give a correct representation of the ether and that there can be no instantaneous propagation of anything in them."[23] If his model indeed gave "a correct representation of the ether," as FitzGerald believed it did, then the equations that had led Maxwell to conclude that the electric

21. These notebooks were held by the Physics Department at Trinity College Dublin until 1989 but have now been moved to the manuscripts room of the main TCD library.
22. Maxwell 1873, art. 783; see also Bork 1966a.
23. FitzGerald notebook 10376, p. 10, FG-TCD Library.

potential was propagated instantaneously must contain some error. FitzGerald set out to find this error and if possible to correct it.

It was at this point that FitzGerald began to sketch arrays of tiny wheels connected by rubber bands and to analyze their motions mathematically. He found that these motions could be decomposed neatly into terms corresponding to the vector and scalar potentials only in the special case of a steady field. When the fields varied, as when electromagnetic waves were passing through a region, the potentials could not adequately represent propagation phenomena as they appeared on the wheel model. "There is nothing exactly like F, G, H [the components of the vector potential **A**] in my wheel system," FitzGerald noted at one point; elsewhere he remarked, "My coordinates αβγ are much nicer ones to use as they represent something in my wheel medium."[24] The vector (α, β, γ) represented the total angle through which the wheels had turned and corresponded to the time integral of the magnetic force; it was equivalent to the vector **R** that FitzGerald had introduced in his 1878 paper on MacCullagh's theory and to the vector **Z,** "the vector potential of the magnetic current," that Heaviside had used in his own theoretical work. The new vector was an important addition to the representation of the electromagnetic field, especially its energy, in terms of potentials; moreover, as Heaviside pointed out, by completing the symmetry of the potential equations, it pointed the way to the simpler scheme in which the potentials were dispensed with and the field treated purely in terms of the electric and magnetic forces.[25]

FitzGerald's analysis of his model eventually led him to the source of Maxwell's difficulties about instantaneous propagation. In deriving his wave equation, Maxwell had introduced an auxiliary quantity J, defined as the divergence of the vector potential: $J = \text{div } \mathbf{A}$.[26] In the modern form of classical field theory, the vector potential is regarded as having no direct physical significance, and since its divergence drops out of the equations connecting **A** to the observable forces, div **A** is simply assigned whatever value is most convenient for the problem at hand. This is called choosing a "gauge," and within broad limits it is a matter of free assumption.[27] But Maxwell regarded the vector potential as far more than a calculational convenience; he called it "the fundamental quantity in the theory of electromagnetism" and invested it with great physical significance.[28] He felt compelled to treat its divergence with similar se-

24. Ibid., pp. 32, 17.
25. Heaviside 1892, 1:466–68 [1885], and 2:484 [1888].
26. Maxwell 1890, 1:578–82 [1864]; cf. Maxwell 1873, arts. 616, 783.
27. For a modern discussion of gauge conditions, see Phillips 1962: 46–48. Some quantum effects depend directly on the vector potential; see Feynman 1964, 2:15.8–15.14.
28. Maxwell 1873, art. 540; cf. Bork 1967.

riousness, and rather than simply assigning a convenient value to J, he sought to deduce its value from physical considerations.

Maxwell was able to show that, in free space,

$$d^2J/dt^2 + d(\nabla^2\psi)/dt = 0.$$

He then made an important assertion, for which he provided no real justification: "$\nabla^2\psi$," he said, "which is proportional to the volume-density of the free electricity, is independent of t"—that is, he claimed that the electric potential is determined solely by the spatial distribution of charge, which in a nonconductor does not change.[29] This was the assumption usually made in electrostatics, and Maxwell simply extended it to general electromagnetic theory without alteration or explanation. Time independence implied that the electric potential adjusted instantaneously across all space to any changes in the positions or magnitudes of the charges; it also implied that $d^2J/dt^2 = 0$, so that "J must be a linear function of t, or a constant, or zero, and we may therefore leave J and ψ out of account in considering wave disturbances."[30] In practice, Maxwell generally took $J = 0$ and so worked in what would now be called the "Coulomb gauge," a gauge well suited to electrostatic problems but with the serious drawback in treating changing fields that it requires the electric potential to be propagated instantaneously.[31]

FitzGerald had already found that there was "no instantaneous propagation of anything" in his wheel model, and in tracing the correspondence between his model and Maxwell's equations he found it impossible to assume, as Maxwell had, that $d(\nabla^2\psi)/dt = 0$ and $J = 0$. He sprinkled the relevant passages in his copy of Maxwell's *Treatise* with question marks and wrote, in the margin, "I'm not so sure of this"; in his notebook, he declared, "The great difficulty is the $dF/dx + dG/dy + dH/dz = 0$," (i.e., div $\mathbf{A} = J = 0$).[32] This condition had no counterpart in the wheel model, and it was here that FitzGerald broke decisively with Maxwell's treatment of propagation. "My wheel motion ought to succeed," he wrote, but only if he dropped Maxwell's assumption that J and ψ were independent and instead put $J = -d\psi/dt$, or, expressed another way, div $\mathbf{A} + d\psi/dt = 0$.[33] This "Lorentz gauge," as it was later called, is much

29. Maxwell 1873, art. 783.

30. Ibid., and art. 616; cf. Bork 1966a: 847–48. The discussion in Buchwald 1985a: 277 implies that Maxwell started by assuming $J = 0$ and was led by this to conclude that ψ is not propagated; in art. 783, Maxwell in fact started by assuming that $\nabla^2\psi$ does not vary with time and derived his conclusions about J from this condition.

31. Phillips 1962: 48.

32. FitzGerald's copy of Maxwell's *Treatise*, note at art. 783, FG-TCD Physics, and FitzGerald notebook 10376, p. 32, FG-TCD Library.

33. FitzGerald notebook 10376, p. 28, FG-TCD Library.

better suited to treating propagation phenomena than was Maxwell's Coulomb gauge, $J = 0$. When combined with Maxwell's equation $d^2J/dt^2 + d(\nabla^2\psi)/dt = 0$, FitzGerald's condition implied that, in free space,

$$\nabla^2\psi - d^2\psi/dt^2 = 0$$

and

$$\nabla^2\mathbf{A} - d^2\mathbf{A}/dt^2 = 0.$$

These are homogeneous wave equations, and they imply that the potentials, including ψ, are propagated in a way completely different from that assumed by Maxwell. Rather than adjusting instantaneously across all space, ψ is now seen to propagate, along with $\mathbf{A}$, $\mathbf{E}$, and $\mathbf{H}$, in waves traveling at the speed of light. The new "gauge" thus eliminated any question of the instantaneous propagation of electric potential.

FitzGerald saw how crucial this change was. During the Bath meeting of the British Association in September 1888, he went back through his notebook and circled his derivation of the new equation for J, scrawling beside it: "very important. 9.9.88. Must be all in O. Heaviside."[34] This was apparently the first direct notice FitzGerald had ever taken of Heaviside's work on propagation, which had in fact already gone well beyond FitzGerald's own, and it foreshadowed the close alliance that later grew up between the two men.

FitzGerald was often rather slow to publish his work, and in the latter part of 1888 and through much of 1889 he was further hampered by recurring digestive troubles and a bout with quinsy. He did not give a public account of his modification of Maxwell's propagation equations until the 1890 Leeds meeting of the British Association, when he presented a short paper "On an Episode in the Life of J," in which he pointed out the source of Maxwell's difficulties and explained the advantages of taking $J = -d\psi/dt$ rather than $J = 0$.[35] He showed how the revised equations could be used to treat electromagnetic waves and other propagation phenomena and pointed out explicitly how the new equations in $\mathbf{A}$ and ψ could be used to calculate the electric and magnetic forces $\mathbf{E}$ and $\mathbf{H}$, which he recognized as the quantities of real physical interest. Indeed, FitzGerald eventually turned away from the potentials altogether and came to denounce their use "an analytical juggle" that obscured the real physics of how energy was localized and propagated in an elec-

34. Ibid., opp. p. 37.
35. FitzGerald 1890. Note a misprint on p. 755; it should read $J = -d\psi/dt$, not $J = d\psi/dt$. This paper was not reprinted in FitzGerald 1902 but is discussed in Lodge 1931a: 136–37.

tromagnetic field.[36] Heaviside had already reached substantially the same conclusion, and perhaps the most important long-term effect of FitzGerald's own analysis of electromagnetic propagation was to make him especially receptive to Heaviside's work when it began to come to his attention in the late 1880s.

"When Energy Goes from Place to Place . . ."

Like FitzGerald, Heaviside was first led to revise Maxwell's theory in an effort to clear up difficulties connected with the propagation of energy. But where FitzGerald had relied on the workings of a mechanical model, Heaviside preferred to follow a more direct and in the end more far-reaching route, sifting and modifying the field equations themselves in search of those that corresponded as closely as possible to the physical state of the field. He was guided throughout by the principle of energy flow, as summarized in one of his characteristic couplets: "When energy goes from place to place, it traverses the intermediate space."[37] By investigating exactly *how* energy goes from place to place in an electromagnetic field and how this flow could best be expressed, Heaviside found the clue to his comprehensive recasting of Maxwell's equations.

Heaviside's interest in the propagation of energy grew quite naturally out of his work in telegraphy. In the final analysis, the important thing that travels along a telegraph wire is energy—the power to deflect a needle or trip a key—and the clean transmission of this energy, without distortion or excessive loss, is the fundamental aim of all telegraphy. Yet before Heaviside took up the problem in the 1880s, very little was really known about how wires guided and controlled the flow of energy in telegraphic circuits.

In 1883, Heaviside began to address these questions in his *Electrician* series, "The Energy of the Electric Current." He followed Maxwell's physical ideas quite closely in these articles but insisted even more strongly on treating electromagnetic energy as distributed throughout the field. He argued on general physical grounds that the energy at a point must depend solely on the conditions at that point, adducing this as a further reason for rejecting the $\frac{1}{2}$ **A·C** formula, since the value of **A** at a point depends on the state of the whole system.[38] He wanted to know as accurately as possible the physical state of the field at each point, and

36. FitzGerald 1902: 509 [1900]; see below, Chap. 9.
37. Heaviside 1893–1912, 1:74 [1891].
38. Heaviside 1892, 1:249 [1883].

he regarded this state as tightly bound up with the location and motion of the energy.

Before 1884, Heaviside shared the usual view that the energy of an electric current flowed along within the wire, much like water in a pipe. Like all field theorists, he believed there was energy in the surrounding field, but he regarded it as having flowed there from the central wire. He described the process early in 1883: "When, then, we set up a current in a conductor, a transmission of energy outwards at once begins, and not until it is completed does the current get steady. . . . Then we have a definite amount of energy in every place where the magnetic force extends, falling off in density of distribution rapidly as we recede from the current. A medium of some kind to receive the energy is, of course, necessary." Following Thomson and Maxwell, Heaviside suggested that this energy might be stored kinetically as the rotation of tiny vortices or "flywheels" in the ether. When the supply of energy from the battery is cut off, he said, the flywheels' "reserve store comes into action [and] is returned to the same wire, or to other conductors in its neighbourhood, creating the phenomena of induced currents."[39] The image here is of a small amount of energy flowing out from the wire into the surrounding medium, while most of the energy simply passes along within the wire; when the current is shut off, the energy in the medium flows back into the wire and is dissipated as heat. Heaviside did not make his account of this process exact or explicit, however, nor did he pursue its implications; he had not yet found a thread with which to tie together his ideas about the energy in the electromagnetic field. Through 1883 and early 1884 his series in the *Electrician* wandered along with articles on Ohm's law, thermoelectricity, and other topics that were sometimes interesting in themselves but yielded no comprehensive view of electromagnetic energy.

The guiding principle Heaviside needed eventually emerged from what began as one of his narrowest and most specialized investigations. In May 1884 he started a new series in the *Electrician*, "The Induction of Currents in Cores," in which he examined what went on in the core (usually iron) of an electromagnet when it was switched on: how were currents induced in it and energy lost in heat, and how was its magnetic field set up? He was interested in how telegraphic receivers, particularly the Wheatstone automatic, responded to rapidly varying signal currents, and he set out to treat mathematically, from what he called "the Maxwellian view," the process by which magnetic force diffuses into a

---

39. Ibid., pp. 249–50 [1883].

core from its surrounding coil, with the concomitant production of heat in both the coil and the core.[40]

This effort led Heaviside quite naturally to an investigation of the "Transmission of Energy into a Conducting Core," published in June 1884. To analyze the rate at which energy entered the core, he used the $\frac{1}{2}\mu H^2$ formula for the magnetic energy density and Joule's well-known law for the generation of heat by currents. His key insight was to focus explicitly on the *flow* of energy: "Now, energy can only enter the portion of core considered across its boundary," he observed, "and, after having entered, is either stored up temporarily as magnetic energy, or is dissipated as heat through induced currents. Hence the rate of passage of energy into the space from outside equals the sum of the rate of increase of magnetic energy and of the rate of dissipation within the space."[41] Heaviside was arguing that electromagnetic energy followed identifiable paths through space and that, once one knew how this energy was distributed and how it was converted into other forms, one should be able to find a formula expressing the rate and direction of its flow. He did not yet know that Poynting had hit upon the same idea more than six months before.

Heaviside's analysis of the energy flux in a conducting core was very complex, and he did not give the details of his derivation in print. His final result, however, was remarkably simple: $\mathbf{S} = r\mathbf{i} \times \mathbf{H}$. That is, the flow of energy is proportional to the vector product of the current and the magnetic force, and thus it is perpendicular to both.[42] This was Heaviside's first statement of the Poynting flux; it takes on a more familiar form when we apply the local form of Ohm's law, $\mathbf{E} = r\mathbf{i}$ (the electric force equals the product of the resistivity and the current density), yielding $\mathbf{S} = \mathbf{E} \times \mathbf{H}$.

Heaviside apparently did not learn of Poynting's paper on the flux theorem until late 1885 or early 1886, a state of affairs that was not at all unusual for him.[43] Working as he did on the fringes of the scientific community, Heaviside had little access to the publications of the Royal Society or, more important, to the informal channels of scientific communication. He often had to piece together major bodies of knowledge from scattered hints in the *Electrician* and the *Philosophical Magazine*, and Poynting's was not the only theorem he rediscovered on his own: he once

40. Ibid., pp. 362–63 [1884].
41. Ibid., p. 377 [1884].
42. Ibid., p. 378 [1884].
43. Heaviside first acknowledged Poynting's priority in an August 1886 *Phil. Mag.* paper, repr. Heaviside 1892, 2:172.

told FitzGerald that when he finally read Fourier's *Analytical Theory of Heat*, he was able "to appreciate its beauty fully at once without the painful process of having to learn it first," since he had already "practically rediscovered all Fourier's mathematics and a lot more too."[44]

Heaviside's frequent ignorance of what others had already accomplished could hardly be called an advantage, but by leaving him free of prevailing prejudices, it sometimes led him to pursue lines of thought that more orthodox scientists and mathematicians might have missed. Had he known and accepted the usual standards of "rigor" in the treatment of infinite series, for example, it is unlikely he would ever have arrived at his "operational calculus," now recognized as his greatest mathematical innovation.[45] In a somewhat similar way, Heaviside's distance from the mainstream of British mathematical physics made it easier for him to dispense with the potentials and the Lagrangian methods favored by members of the "Cambridge school" and to strike out in new and, as he believed, more fruitful directions.

Heaviside's Equations

The summer of 1884 was an exciting time for Heaviside. With the discovery of the flux formula, he at last had a firm physical principle to help guide him through the intricacies of electromagnetic theory, and once he began to focus on the flow of energy and the primacy of **E** and **H,** the pieces of the puzzle quickly fell into place. "After that I did more in a week than in all the previous years," he later told Hertz; "in fact, I sketched out all my later work."[46] In particular, it was in this period that he recast the basic equations of Maxwell's theory into a new and much simpler form—into what might properly be called "Heaviside's equations." By moving to this now-familiar set of four vector equations, Heaviside was able to make Maxwell's theory far more concise and workable; more important, in his view, he was able to bring it into harmony with basic dynamical principles.

Like most of his British contemporaries, Heaviside learned dynamics from "that stiff but thorough-going work," Thomson and Tait's 1867 *Treatise on Natural Philosophy.*[47] But while most other mathematical physicists, particularly those of the Cambridge school, looked to Thomson

44. Heaviside to FitzGerald, 26 Feb. 1894, FG-RDS.
45. See Nahin 1988: 217–40 and Hunt 1991.
46. Heaviside to Hertz, 13 July 1889, HH-DM.
47. Heaviside 1893–1912, 3:178 [1903].

and Tait for instruction on Lagrangian methods and the principle of least action, Heaviside focused instead on their treatment of the "principle of activity,"[48] which states that the rate at which a force adds energy to a system is the product of the applied force and the velocity of the point at which it acts. Thus a force $\mathbf{F}$ acting on a mass m moving with velocity $\mathbf{v}$ will, as it accelerates the particle, add energy at the rate $\mathbf{F} \cdot \mathbf{v} = m\dot{\mathbf{v}} \cdot \mathbf{v}$, which is the rate at which the kinetic energy, $T = \frac{1}{2} mv^2$, increases. Heaviside regarded the principle of activity as the best guide to local transformations of energy and so to the real workings of dynamical systems, and his reliance on that principle set his work somewhat apart from that of more conventional British physicists.[49]

In the summer of 1884, Heaviside asked himself how the principle of activity might be applied to Maxwell's theory. If the electromagnetic field is indeed the "connected dynamical system" Maxwell always insisted it was, then one should be able to find the energy flow vector simply by calculating the rate at which the electric and magnetic forces act on the corresponding "velocities" (the electric and magnetic currents). But in fact, Heaviside said, "If we take Maxwell's equations [as given in his *Treatise*] and endeavour to immediately form the equation of activity (like $Fv = \dot{T}$ from $F = m\dot{v}$), it will be found to be impossible. They will not work in the manner proposed." The energy flow could be derived from Maxwell's original equations only by a long and indirect route: "We may," Heaviside said, "consider the energy, electric and magnetic, entering and leaving a given space, and that dissipated within it, and by laborious transformations evolve the expressions for the vector transfer."[50] Poynting had used this approach, as had Heaviside himself in his initial investigation in June 1884. But Heaviside was dissatisfied (as Poynting apparently was not) with "the roundabout nature of the process to obtain what ought to follow immediately from the equations of motion." Convinced that there must be an easier way, he decided to work backward from the flux formula to find what the basic equations of motion ought to be. This led him, he said,

to remodel Maxwell's equations in some important particulars . . . with the result of producing important simplifications, and bringing to immediate view useful analogies which are in Maxwell's equations hidden from sight by the intervention of his vector potential. This done, the equation of ac-

48. Thomson and Tait 1867: 72. On "Thomson and Tait" and its influence on British physics, see Smith and Wise 1989: 348–95.
49. See Buchwald 1985b: 228 and Buchwald 1985c.
50. Heaviside 1892, 2:92–93 [1887].

tivity is at once derivable from the two cross-connections of electric force and magnetic current, magnetic force and electric current, in a manner analogous to Fv = $\dot{T}$, without roundabout work.[51]

Heaviside's procedure can be followed fairly easily in the simple case of free space. With **E** and **H** as the electric and magnetic forces and $\dot{\mathbf{D}}$ = $\epsilon\dot{\mathbf{E}}$ and $\dot{\mathbf{B}}$ = $\mu\dot{\mathbf{H}}$ as the corresponding currents, the equation of activity yields

$$\dot{W} = \mathbf{E}\cdot\epsilon\dot{\mathbf{E}} + \mathbf{H}\cdot\mu\dot{\mathbf{H}}$$

as the rate of increase of the energy density, W. Maxwell had given $\epsilon\dot{\mathbf{E}}$ = curl **H** in his *Treatise,* and by combining two equations involving the potentials **A** and ψ, Heaviside was able to derive $\mu\dot{\mathbf{H}}$ = −curl **E**. He thus obtained

$$\dot{W} = \mathbf{E}\cdot(\text{curl } \mathbf{H}) - \mathbf{H}\cdot(\text{curl } \mathbf{E}),$$

which by the rules of vector algebra is equivalent to

$$\dot{W} = -\text{div} (\mathbf{E} \times \mathbf{H}).$$

The rate of increase of the energy density is thus equal to the convergence of **E** × **H**, demonstrating that the Poynting vector is indeed the energy flux.[52]

The key step was the derivation of −curl **E** = $\mu\dot{\mathbf{H}}$ (see Appendix). This equation enabled Heaviside to dispense with the potentials and to focus entirely on the electric and magnetic forces and fluxes. He was thus led to "an entirely new way of developing Maxwell's electromagnetic scheme, with a new fundamental equation, −curl E = $\mu\dot{\mathbf{H}}$, abolishing ψ and A altogether, and working with E and H from the beginning."[53] The new equation relating the rate of change of the magnetic flux to the curl of the electric force was simply a direct mathematical statement of Faraday's law of electromagnetic induction, without the circumlocutions introduced by Maxwell's use of the vector potential. Heaviside emphasized the parallel between this "second circuital law" and Maxwell's "first cir-

51. Ibid., p. 93.
52. See Feynman 1964, 2:27.3–27.5. Any flux **G** such that div **G** = 0 can, of course, be added to **S** without changing the net result through any closed surface; cf. Nahin 1988: 131.
53. Heaviside notebook 3a, entry 78, OH-IEE.

cuital law" relating the electric current to the curl of the magnetic force; together they formed a pair of symmetrical "duplex equations." This formal symmetry reflected a deeper physical symmetry, Heaviside said, and he regarded the fact that the activity and energy flow could be derived directly from these equations, without detouring through the potentials, as proof that his second circuital law was "really the proper and natural fundamental equation to use."[54]

The potentials, according to Heaviside, were vestiges of an outmoded action-at-a-distance approach to electromagnetism; they were "functions considerably remote from the vectors which represent the state of the field" and had no place in the fundamental equations of Maxwell's theory.[55] Once they were dropped, he told Hertz in 1889, "a heap of metaphysics disappears. Besides that, problems can now be readily attacked that are extremely obscure and unmanageable when expressed in terms of potentials from the beginning. Boundary conditions are made self-evident, and work is generally much lightened. Important properties are recognised which were previously hidden from view by the intervention of the potentials. Etc. etc."[56] In short, Heaviside found that eliminating the potentials enabled him to cast Maxwell's theory into a particularly simple and useful form. Instead of the cumbersome tangle of equations given in the *Treatise*, he was left with the concise and elegant set of four now universally known as "Maxwell's equations." For free space, these are:

$$\text{div } \epsilon\mathbf{E} = 0 \qquad \text{curl } \mathbf{H} = \epsilon\dot{\mathbf{E}}$$
$$\text{div } \mu\mathbf{H} = 0 \qquad -\text{curl } \mathbf{E} = \mu\dot{\mathbf{H}}.$$

This remarkably simple and symmetrical set of equations suffices for the treatment of all purely field phenomena—in particular, for the propagation of electromagnetic waves. Their formulation marked a milestone in the development of Maxwellian theory.

The equations become somewhat more complicated when electric charges and conduction currents are present. With ρ as the charge density and k as the electric conductivity, the upper two equations become:

$$\text{div } \epsilon\mathbf{E} = \rho$$

54. Heaviside 1892, 2:174 [1886].
55. Heaviside 1893–1912, 1:127 [1891].
56. Heaviside to Hertz, 14 Feb. 1889, HH-DM.

and

$$\text{curl } \mathbf{H} = k\mathbf{E} + \epsilon\dot{\mathbf{E}}.$$

The divergence of the electric displacement through a closed surface gives the enclosed charge, while the total current, as measured by the curl of the magnetic force, is the sum of the displacement current and the conduction current.[57] To preserve symmetry, Heaviside suggested that terms for magnetic charge and magnetic conduction current should also be introduced.[58] With σ as the magnetic charge density and g as the magnetic conductivity, the two lower equations become:

$$\text{div } \mu\mathbf{H} = \sigma$$

and

$$-\text{curl } \mathbf{E} = g\mathbf{H} + \mu\dot{\mathbf{H}}.$$

There was no evidence that either magnetic monopoles or magnetic conduction currents really existed, but maintaining the symmetry of the equations had important analytical advantages, notably in the treatment of propagation along wires (as we will see in the next chapter). In practice, Heaviside simply took σ and g to be zero.

Heaviside's "duplex" equations were not entirely unknown before he wrote them down in 1884. Maxwell had in fact stated them himself (in words rather than symbols) in his brief "Note on the Electromagnetic Theory of Light" in 1868. He gave a clear statement of the second circuital law relating the magnetic current to the curl of the electric force, crediting it to Faraday and asserting that it afforded "the simplest and most comprehensive" expression of the facts.[59] But despite the advantages of this simple equation, particularly in treating electromagnetic waves, Maxwell did not use it at all in his *Treatise,* preferring to work entirely with the vector potential. His 1868 "Note" remained unknown even to close students of his theory until after his *Scientific Papers* were published in 1890, and the fact that the second circuital law could be

57. Moving charges give rise to a third term, the convection current ρv; see Heaviside 1892, 2:492–93 [1888].
58. On magnetic conductivity, see Heaviside 1892, 1:441–43 [1885]; on magnetic charge ("magnetification"), see Heaviside 1893–1912, 1:50–51 [1891]. See also Hendry 1983.
59. Maxwell 1890, 2:138–39 [1868].

found in Maxwell's own writings was news to Lodge as late as 1899.[60] Rayleigh independently derived the second circuital equation in 1881, and Hertz, working from a generalized action-at-a-distance theory, arrived at a set of symmetrical equations similar to Heaviside's in 1884; but neither of these derivations had any discernible effect on the way Maxwell's theory was understood and used.[61] Heaviside was the first to assert that the symmetrical set of equations should be regarded as fundamental, and it was mainly through his efforts that the new form of "Maxwell's equations" eventually came into general circulation. He campaigned strongly for the adoption of the new equations, first in the long series on "Electromagnetic Induction and Its Propagation," which he began in the *Electrician* in January 1885, and later in papers published in the *Philosophical Magazine* between 1886 and 1888.[62] He made cogent general arguments for the superiority of his equations, but more important, he showed how they could be used to solve a wide range of practical and theoretical problems. Heaviside had tailored his notation, his vector algebra, and now his form of Maxwell's equations to fit as closely as possible his object of study: the electromagnetic field. The result was a set of mathematical tools uniquely suited to the solution of electromagnetic problems, particularly those involving propagation, and the manifest power and utility of Heaviside's methods soon won others over to their use. By the mid-1890s his equations were widely regarded as the standard form of Maxwell's theory and had begun to take their place in an influential new generation of textbooks.[63] The potentials came to be viewed as little more than analytical expedients, and few readers of twentieth-century textbooks of electromagnetism would ever guess that Maxwell had built his theory on $\mathbf{A}$ and ψ.

Heaviside had only "redressed" Maxwell's theory, not changed its basic content. Except on a few points of detail, his new formulation was mathematically equivalent to that given in the *Treatise*. But Heaviside's reforms were by no means trivial; as he told Lodge in 1888, they were, "to the initiated, of the greatest importance, involving a complete

60. Larmor was apparently the first to draw attention to Maxwell's 1868 statement of the second circuital law; see Larmor 1929, 1:555n [25 Aug. 1895]. In his presidential address to the Physical Society in Feb. 1899, Lodge noted that the second circuital law "is now largely adopted and greatly simplifies Maxwell's treatment" and said that its formulation was "entirely due to Mr. Heaviside." Someone (presumably Larmor) then alerted him to Maxwell's 1868 paper, prompting Lodge to add a correction before publication; see Lodge 1899: 372n, 386 [May 1899].

61. Rayleigh 1881; Hertz 1896: 273–90 [1884]. See also D'Agostino 1975: 286–91.

62. Heaviside 1892, 1:429–560, 2:39–155 [1885–87], and 2:168–467 [1886–88].

63. One of the best and most influential was Föppl 1894, which followed Heaviside's methods quite closely; see the praise of it in Heaviside 1893–1912, 3:504 [1897].

change of procedure."[64] His new notation and equations embodied a set of techniques and concepts that differed radically from those associated with the use of the potentials. By making the forces and energy in the field the direct objects of theoretical attention, Heaviside helped bring the formal expression of Maxwell's theory into closer accord with its underlying physical principles.

In an 1893 review for the *Electrician* of Heaviside's *Electrical Papers,* FitzGerald declared, "Since Oliver Heaviside has written, the whole subject of electromagnetism has been remodelled by his work." Maxwell had laid down the principal ideas of field theory, FitzGerald said, but had been unable to follow out all their implications:

> Maxwell, like every other pioneer who does not live to explore the country he opened out, had not had time to investigate the most direct means of access to the country, or the most systematic way of exploring it. This has been reserved for Oliver Heaviside to do. Maxwell's treatise is cumbered with the *débris* of his brilliant lines of assault, of his entrenched camps, of his battles. Oliver Heaviside has cleared those away, has opened up a direct route, has made a broad road, and has explored a considerable tract of country.[65]

When the 1880s began, Maxwell's theory was virtually a trackless jungle. By the second half of the decade, guided by the principle of energy flow, Poynting, FitzGerald, and above all Heaviside had succeeded in taming and pruning that jungle and in rendering it almost civilized.

64. Heaviside to Lodge, 9 Oct. 1888, OJL-UCL.
65. FitzGerald 1902: 299 [1893].

CHAPTER SIX

Waves on Wires

Between 1885 and 1887, Heaviside began to use his "redressed" version of Maxwell's equations to investigate how electrical signals travel along wires. Attempts to extend telephone lines to long distances were beginning to encounter problems with distortion, and the whole issue of propagation along wires was becoming a matter of urgent interest. Guided by a characteristic Maxwellian belief in the primacy of the field, and especially by his newly discovered energy flux theorem, Heaviside came to conceive of electrical signals in a fundamentally new way, not as pulses of current within the wire itself but as trains of electromagnetic waves in the surrounding dielectric. His analysis of how such waves lose energy to the wires along which they run led him in 1887 to the important discovery that distortion could be dramatically reduced, even eliminated, simply by loading a line with extra self-induction.

Inductive loading eventually proved of enormous value in long-distance telephony, but the practical application of Heaviside's discovery was delayed for more than a decade. W. H. Preece and other "practical men" resented this attempt by an obscure theoretician to tell them how to build their telephone lines, and Heaviside's claims about waves on wires and inductive loading became the focus of a heated battle between "practice" and "theory" which split the British electrical community in the late 1880s.[1] Heaviside was blocked for a time from publishing some of his most important work, and it appeared that he might never win a proper hearing for his ideas. He fought back by deliberately seeking the

1. See Jordan 1982a and Hunt 1983.

attention and support of leading scientists, but his campaign for recognition would probably have failed had it not been for the unexpected experimental discovery of electromagnetic waves in 1888 by Oliver Lodge and Heinrich Hertz. In the wake of their discoveries, Heaviside's theoretical work on electromagnetic propagation became the focus of intense scientific interest, and Heaviside himself was finally drawn more fully into the scientific community. The dispute over waves on wires marked a turning point in Heaviside's own career, and it played a major role in the development of the Maxwellian group as a whole.

"Beams of Dark Light"

"The question of the function of wires is the most important one in all of electromagnetism," Heaviside declared in 1888, yet it long remained one of the least understood.[2] The seemingly natural idea that electrical energy and signals flow within a wire, much like water in a pipe, dominated almost all work on electromagnetic propagation before the mid-1880s. Even William Thomson, despite the strong influence of Faraday's field ideas, had continued in his 1855 cable theory to treat signals as traveling *in* the conducting wire.[3] This worked well enough as long as the theory was applied only to the high-capacitance submarine cables for which Thomson had first devised it, since in them the propagation of current and potential was almost indistinguishable from simple diffusion along the conductor. But over the more than thirty years that Thomson's theory served as the basis for the mathematical treatment of telegraphic propagation, many less cautious electricians applied it well outside its proper range. Especially widely misapplied was Thomson's conclusion that the rate of signaling was limited solely by the product of the total capacitance (K) and resistance (R) of the line. Preece in particular applied this simple "KR Law" very broadly, extending it even to overhead telephone lines.[4] The propagation conditions on such low-capacitance lines are very different from those on submarine cables, especially for high-frequency telephonic signals, and it is not surprising that the KR law proved a poor guide to efficient design. Although telephone engineers continued to invoke the KR law in one form or another for many years, there were already signs in the 1880s that it would

2. Heaviside to Lodge, 17 Sept. 1888, OJL-UCL.
3. William Thomson 1882–1911, 2: 61–76 [1855]; cf. Smith and Wise 1989: 446–53.
4. Preece 1887d; see Nahin 1988: 147, 174, 177. Preece did not coin the name "KR Law" until 1887, but the idea behind it was by then already well known.

eventually have to be replaced by a new and more general theory of propagation along wires.

When Heaviside first took up the question of telegraphic propagation in the 1870s, he simply applied Thomson's theory to various special cases he had encountered on the Anglo-Danish cable. He soon found the simple 1855 theory inadequate and began to extend it to take account not only of resistance (R) and capacitance (for which Heaviside used the symbol S) but also of inductance (L) and leakage conductance (for which he used the symbol K). The result was a very sophisticated version of the water-in-a-pipe theory, with a much broader range of applicability than Thomson's 1855 theory but with most of the same underlying physical ideas and the same focus on the current within the wire itself. Heaviside was able to extract many useful results from his four-parameter cable theory, but the mathematical obstacles to its complete understanding and application were enormous, and the physical image of electricity flowing in a wire provided little useful guidance. It was only after he turned to Maxwell's field theory in the 1880s that Heaviside found the new physical ideas that enabled him to go beyond the old water-in-a-pipe view and, as he later said, "to let in the daylight on a subject which it was difficult to believe could ever be freed from mathematical complications."[5]

The essence of the new Maxwellian approach to propagation along wires was stated very succinctly by FitzGerald in 1885. In his October 1884 Baltimore lectures, Thomson had repeated his claim, first made in the 1850s, that electrical signals move through a submarine cable much like water in a rubber tube, the elasticity of the tube corresponding to the capacitance of the cable.[6] FitzGerald objected that taking account only of a thin rubber tube was insufficient; to make even a crude analogy to the real situation, the space around the conductor would have to be pictured as completely filled with rubber, he said, and careful attention paid to the propagation of waves and energy within this elastic "field." FitzGerald called for a complete shift of focus. "According to Maxwell's view," he stressed, "there is a great deal more going on outside the conductor than inside it."[7]

Heaviside took substantially the same view and had already begun to develop its consequences in considerable detail. He wanted to eliminate altogether the idea of a literal electric "current," and he expressed a

5. Heaviside 1892, 2: 134 [1887].
6. William Thomson, "Lecture IV" [1884], in Kargon and Achinstein 1987: 42; cf. Forbes 1885: 463.
7. FitzGerald 1902: 172 [1885].

hope that the new Maxwellian view of propagation along wires would "assist in abolishing the time-honoured but (in my opinion) essentially vicious practice of associating the electric current in a wire with the motion through the wire of a hypothetical *quasi*-substance, which is a pure invention that may well be dispensed with."[8] A wire should be viewed, he said, not as a conduit within which electromagnetic waves and energy flow but as no more than the relatively quiescent core of a disturbance in the ether around it.

In this view, the main function of a conducting wire is to enable part of the accumulating electrical strain in the surrounding field to break down, thus preventing the electromagnetic machinery in the ether from jamming itself as it does in a condenser as it becomes fully charged. In and around a charged condenser, a substantial amount of energy is stored in the ether in a self-locked state of strain; in a leaky condenser or a conducting wire, this strain is continually breaking down, and energy and strain have to flow through the ether to make up the difference.[9] By providing a dumping ground for part of the strain in the field, the wire keeps the lines of magnetic force wrapped tightly around itself and serves to guide the main flow of energy from one place to another. High-frequency signals of the kind used in telephony and high-speed telegraphy barely enter the wire at all, Heaviside said, but instead slip along its outer surface "like greased lightning."[10] They travel essentially as electromagnetic waves in the surrounding ether, and it was one of Heaviside's most penetrating insights to realize that "telegraph wires made beams of dark light."[11] This new conception of propagation along wires, in which the focus was firmly on the flow of waves and energy in the surrounding electromagnetic field, armed Heaviside with a powerful set of physical ideas which he proceeded to bring to bear on some of the main practical problems of telegraphy and telephony.

Loading and the Distortionless Circuit

Heaviside's new approach to electromagnetic propagation played a key role in his discovery of the ideal distortionless circuit. Such a circuit,

8. Heaviside 1892, 2: 143 [1887].
9. Buchwald 1985a: 23–40 gives a very full account of the Maxwellians' concepts of charge and current, but his contention that they regarded conduction as the result of an *intermittent* decay of dielectric displacement seems to me to be based not on Maxwellian theory itself but only on some of the speculative efforts to give it a molecular basis.
10. Heaviside 1892, 2: 140 [1887].
11. Heaviside to Lodge, 5 Jan. 1899, OJL-UCL. Lodge took up this phrase in Lodge 1899: 368.

in which the electrical parameters are balanced in such a way that waves pass along it wholly without distortion, provided a "Royal Road," Heaviside said, to the proper understanding of electromagnetic waves and telegraphic propagation.[12] Besides casting a strong light on the physical processes involved in propagation along wires, the distortionless circuit pointed directly toward valuable practical improvements, most notably the inductive loading of telephone lines.

Heaviside's work on distortionless propagation grew in part out of his exploration of a very speculative corner of electromagnetic theory. When he formulated his new "duplex" form of Maxwell's equations in 1885, he had included a hypothetical magnetic conductivity, corresponding to the known electrical conductivity. He had done this, he said, "merely to complete the analogy between the electric and magnetic sides of electromagnetism"; there was no sign in magnetic circuits of resistive heating from the breakdown of magnetic induction or of any other effects analogous to those arising from electrical conduction, and he admitted "There is probably no such thing as a magnetic conduction current, with dissipation of energy."[13] But the inclusion of magnetic conductivity made Heaviside's equations more general and symmetrical and enabled him to bring out more clearly the parallels between electrical and magnetic phenomena. When applying the equations to real situations, he simply took the magnetic conductivity as zero.

Heaviside subsequently used his new equations to investigate how electromagnetic waves would propagate in a slightly conductive medium, and to keep the problem as simple and symmetrical as possible, he assumed the medium to be both electrically and magnetically conductive. As in all his work after 1884, he paid close attention to the flow and dissipation of energy in the medium. He had already shown—indeed, Maxwell had shown—that when the electrical and magnetic energies of a wave are equal, it moves along without distortion, as a light wave does in empty space.[14] The oscillations of the electric and magnetic forces exactly regenerate each other in each cycle, leaving nothing behind. But in an electrically conductive medium, such as leaky insulation, the electric displacement of the wave is partly broken down with each oscillation, and its electrical energy is gradually dissipated as heat. The wave thus comes to have more magnetic than electrical energy, and as it rids itself of excess magnetic induction, it throws off a "tail" and becomes distorted. Heaviside saw that his hypothetical magnetic conductivity could

---

12. Heaviside 1893–1912, 1: 3 [1891], and Heaviside 1892, 2: 340.
13. Heaviside 1892, 2: 379 [1888] and 1: 441 [1885]; cf. Hendry 1983: 82.
14. Maxwell 1873, art. 792.

counteract this process: if the magnetic and electrical conductivities were adjusted so that each extracted energy from the wave at exactly the same rate, the electrical and magnetic energies of the wave would remain in balance, and the wave would move forward without distortion.[15] It would be attenuated by the two conductivities, of course, but retain its shape unchanged.

This was all very interesting but not of much apparent use: magnetic conductivity was, after all, purely hypothetical, and electrical conductivity acting alone simply caused distortion. Late in 1886, however, while analyzing a new kind of telephone circuit his brother Arthur had devised, Heaviside found that the ordinary electrical conductivity of a wire could imitate the effect of the imaginary magnetic conductivity of the dielectric, making virtually distortionless propagation a practical possibility.[16] When an electromagnetic wave runs along a wire, its moving magnetic field induces conduction currents in the wire; as these are dissipated by the electrical resistance of the wire, magnetic energy is extracted from the wave much as it would be by magnetic conductivity in the dielectric.[17] (The parallel is not quite exact, since the point of dissipation is in the wire rather than in the wave itself, but this introduces only a very small correction.) If the loss of energy in the wire were made to occur at exactly the same rate as that in the surrounding dielectric owing to leakage conductance (as it could be, by making L/R, the time constant for the decay of the magnetic energy, equal to S/K, the time constant for the decay of the electric energy), then just as in the imaginary case of magnetic conductivity, the electrical and magnetic energies of the wave would remain in balance, and the wave, though attenuated, would propagate without distortion.

Heaviside said later that it was "only by the introduction of the idea of magnetic resistance (real or virtual) that the problem of propagation along wires can come into Maxwell's theory at all"; only by focusing on the dissipation of magnetic energy by resistance could the theory of electromagnetic waves be brought to bear on telegraphic problems and the nature, cause, and cure of distortion be discovered.[18] The idea of distortionless propagation enabled Heaviside to extract basic physical insights from Maxwell's theory and to translate them into terms appropriate to ordinary telegraphic problems, in which conditions were expressed not as forces and fluxes but as voltages and currents. Building a bridge be-

15. Heaviside 1892, 2: 378–79 [1888].
16. Ibid., p. 402 [1888], and Heaviside 1893–1912, 1: 376–77, 435–37 [1893].
17. Heaviside 1892, 2: 379 [1888] and 2: 512 [1889].
18. Heaviside to Lodge, 2 Jan. 1899, OJL-UCL.

tween field theory and the four-parameter line theory was one of Heaviside's most important and useful achievements. That the condition for truly distortionless propagation was as simple as $L/R = S/K$ must have come as a surprise, but its discovery provided a striking example of how even seemingly arcane theoretical speculations about waves on wires could produce results of great practical value.

As Heaviside was well aware, most telegraph and telephone lines were far from meeting the distortionless condition; their ratio of inductance to resistance (L/R) was generally much less than their ratio of capacitance to leakage conductance (S/K). The discrepancy was greatest in submarine cables, which had very high capacitance because of the iron sheathing and sea water that surrounded them and very low leakage because of the excellence of their gutta-percha insulation. Overhead lines had lower capacitance and higher leakage and so came closer to the ideal, but even they were subject to appreciable distortion when extended to very long distances. The resistance of a line or cable could be decreased only at great expense in extra copper; the capacitance could generally be reduced only slightly and at even greater expense; and a large increase in the leakage would produce so much attenuation that signals would be undetectable over long distances. This left the inductance, which was fairly low on most lines, as the most freely adjustable parameter, and Heaviside's advice was simple: "Increase the leakage as much as is consistent with other things [i.e., to the limit of acceptable attenuation], and increase the inductance greatly."[19] He explained that "loading" a line with extra inductance, distributed as evenly as possible along its length, would increase the magnetic energy of the waves running along it and give them more "electromagnetic momentum" to help carry them along against frictional resistance. Loading thus decreased the attenuation of the waves (counteracting the effect of the increased leakage) and greatly reduced their distortion. The effect was similar to that of distributing birdshot or other small masses along the length of a stretched string to improve its ability to carry transverse waves, and Heaviside later joked that his name and loading were "naturally and providentially connected. You heavify a line by the process of heavification."[20]

Although it was often not practicable to attain truly distortionless propagation without introducing excessive leakage, Heaviside showed theoretically that the addition of even a modest amount of extra induc-

19. Heaviside 1892, 2: 124 [1887].
20. Heaviside to Lodge, 18 Dec. 1918, OJL-UCL; this remark was probably aimed at those who referred to loading as "pupinization," after Michael Pupin. On the analogy to the mechanical loading of strings, see Heaviside 1893–1912, 1: 429–32 [1893].

tance could dramatically reduce the distortion and attenuation on a line and greatly increase the distance over which telephony was possible. He even suggested practical and relatively inexpensive ways to increase the inductance of wires. In 1887 he proposed that finely divided iron be mixed into the insulation of a line to make what he later called an "ironic" conductor, and in 1893 he suggested that discrete high-inductance coils, if spaced closely enough along a line, might serve just as well and be more practicable, as indeed later proved to be the case.[21]

Heaviside planned to announce his discovery of inductive loading in the spring of 1887 in a major paper, "The Bridge System of Telephony," to be written jointly with his brother Arthur.[22] Arthur West Heaviside served as chief engineer for the Northeast District of the Post Office Telegraphs and was one of the pioneers of telephony in Britain. After encountering great problems with distortion when telephones on an underground cable were connected in series, as was then the usual practice, he had hit on the idea of arranging them instead in parallel, as "bridges" between the two main wires of the circuit. This new system greatly improved the quality of transmission, and toward the end of 1886, Arthur obtained permission from his Post Office superiors to write up an account of it for the *Journal of the Society of Telegraph-Engineers and Electricians,* to which Oliver was to contribute a theoretical analysis.

That a circuit on which telephones were connected in parallel should convey signals better than one on which they were connected in series made perfect sense even on the old *KR* theory, since putting the telephones in bridge removed large sources of resistance and capacitance from the circuit. But on applying his new Maxwellian theory of propagation along wires, Heaviside found to his surprise that a circuit with telephones in bridge would be superior even to one on which there was no apparatus at all. It was in following up this clue that he first realized that truly distortionless propagation could be achieved by adding extra inductance and leakage to a line.[23] Both he and Arthur recognized the importance of this discovery, and when they completed their paper in April 1887, they were confident it would be welcomed by both scientists and engineers. They were wrong. Their paper was blocked before it could even appear in print, and its suppression set off a chain of events that had a profound effect on Heaviside and on the British electrical community as a whole.

21. Heaviside 1892, 2: 123 [1887], and Heaviside 1893–1912, 1: 441–45 [1893]; cf. Brittain 1970.
22. Heaviside 1893–1912, 1: 433–37 [1893]; cf. Heaviside 1892, 2: 323n.
23. See Heaviside 1893–1912, 1: 433–37 [1893]. Heaviside first found how to eliminate reflection from the bridges along the circuit and then realized that this condition was closely related to the distortionless propagation of waves.

Suppression

Heaviside's proposal for inductive loading is now recognized as the most important technical innovation in telephone transmission between Bell's original invention in 1876 and the development of the first electronic amplifiers in 1912.[24] Loading greatly improved the quality of long-distance transmission at very modest cost, and when loading coils were finally introduced commercially around 1900 in America, their success helped make a number of scientists, engineers, and corporations—conspicuously not including Heaviside—very rich.[25] There is thus a special irony in the fact that when Heaviside first proposed inductive loading in 1887, British telephone engineers rejected his suggestion outright. In retrospect, inductive loading appears as a prime example of the contribution Maxwell's theory could make to electrical technology, but in the late 1880s not everyone admitted that "mere theory" had anything useful to contribute to practice, and Heaviside's proposal became the focus of a bitter confrontation between scientists and "practical men."

When Heaviside first suggested adding extra inductance to telephone lines, the idea struck most telephone engineers as absurd. They knew from experience that high self-induction in a receiver tended to "choke" high-frequency signals, and it seemed obvious to them that increasing the inductance of an entire line would render it virtually useless for telephony. Preece was especially outspoken about the evils of self-induction, and as chief of the Post Office telegraph engineers, his views carried considerable weight. In a series of papers on the subject in 1886 and 1887 he declared, "The effects of self-induction are invariably ill effects"; self-induction was, he said, "a *bête noir*" and should be eliminated from telephone lines wherever possible. He bolstered this conclusion by invoking Thomson's "*KR* Law," which he claimed applied "precisely [and] in every respect" to telephone lines, and with experiments showing that long-distance speaking quality was best over copper wires whose self-induction was, according to Preece, virtually nil.[26]

Preece did not undertake these extensive and costly investigations out of mere curiosity but as a way to support his campaign within the Post Office to replace iron telegraph and telephone wires with copper ones.

24. Brittain 1970: 36; Wasserman 1985 describes more fully the application of loading by AT&T.

25. Michael Pupin, the American patent holder, reaped profits of about half a million dollars; AT&T, far more; see Brittain 1970: 54 and the interesting but unreliable autobiographical account in Pupin 1924: 330–41.

26. See Preece's remarks in "Discussion of Hughes Address," *JSTEE 15* (1886): 55, 61, and his papers at the 1887 British Association meeting: Preece 1887a, 1887b, and 1887c.

There was considerable evidence that copper wires were superior even to thick iron wires of comparably low resistance, but to make a convincing case for such an expensive change, Preece believed he needed an explanation for this superiority. He thought he had found it in the low inductance of copper circuits, and he made the need to eliminate self-induction the cornerstone of his advocacy of copper wire.[27]

Preece's pronouncements on telephonic theory were widely accepted, but they did not go wholly unchallenged. When he repeated some of them at a meeting of the Society of Telegraph-Engineers and Electricians early in 1887, two prominent London electrical engineering professors, Silvanus Thompson and W. E. Ayrton, responded by denouncing his whole approach to telephonic transmission as absurd.[28] Although Thompson and Ayrton did not realize that self-induction could actually be beneficial, they were sure that Preece's treatment of it was erroneous. They pointed out that his use of the *KR* law was based on the palpably false assumption that Thomson's 1855 cable theory could be applied to overhead telephone lines, an error that vitiated all of Preece's conclusions. Thompson was especially biting at the meeting of 23 February, poking fun at Preece's combination of "cocksureness" and ignorance and "creating a good deal of amusement among the members present," according to the *Electrician* report.[29] Preece was stung by these attacks, and his irritation was compounded when a table of experimental data supporting his use of the *KR* law was inadvertently left out of the issue of the society's *Journal* in which the debate was reported.[30] He responded by digging in, and in March 1887 he presented a paper at the Royal Society "On the Limiting Distance of Speech by Telephone," in which he committed himself even more strongly than before to the *KR* law and to the elimination of self-induction from telephone lines.[31]

In the wake of these events, there could hardly have been a worse time than April 1887 at which to confront Preece with a proposal to add *extra* inductance to telephone lines, but it was precisely then that the Heaviside brothers' paper advocating such inductive loading reached his desk. Although Arthur had obtained permission from his superiors before beginning work on the paper, the final product had to be checked again by Post Office authorities before it could be cleared for publica-

27. Jordan 1982a: 441–42; Preece 1885.
28. Thompson 1887; Preece took a leading role in the first two discussion sessions and was answered by Ayrton and Thompson in the third.
29. *Elec. 18* (1887): 380–81.
30. The table appeared in the next issue with an explanation of the oversight; *JSTEE 16* (1887): 265–68.
31. Preece 1887d.

tion.[32] As chief electrician, Preece was the "official censor" in such cases, and given the circumstances, his reaction was perhaps understandable. He may have sincerely believed his use of the *KR* law and his views on inductance to be incontrovertibly true; in any case, he found opposition to them inconvenient. Although the Heavisides did not mention Preece by name in their paper, the implications of their argument were clear, and Preece was in no mood to let such attacks on his position be made public. When the paper reached him, Preece "fell foul of it in a savage and even insulting manner," according to Heaviside, and refused to allow publication.[33] Preece treated Heaviside's own theoretical sections especially roughly: "P. went through my work," Heaviside said, "marginally annotating it 'Of course,' 'Not new,' 'Too elementary,' and so on, striking out whole paragraphs, and when he came to the end, *ordered* it to be all omitted on the grounds of 'Irrelevancy' and 'Want of novelty.' This order was repeated in a separate document in which he spoke contemptuously of me."[34] Heaviside was outraged; always sensitive to perceived slights, the experience of "being sat upon by an impertinent jack-in-a-telegraph-office" engendered in him a hatred of Preece that never really abated. He made it his personal mission to expose Preece as "a scientific pretender" and a scoundrel, and even after Preece's death in 1913 he continued to denounce him as "an intensely greedy, grasping man" who had stolen and mangled the work of others.[35] Heaviside's hostility at times verged on paranoia, and he came to see Preece behind virtually every misfortune that befell him. In the summer and fall of 1887 those misfortunes were numerous.

The Heavisides sent Preece a strong protest in May, but he still refused to clear the paper. Arthur eventually acquiesced. "He understands his subordinate position," Oliver said, and did not want to jeopardize his career.[36] But the younger Heaviside refused to let the matter drop. At the end of May he sent an account of Preece's actions to the *Electrician*, but the editor, C. H. W. Biggs, who had been very supportive of Heav-

32. The story is told most fully in Heaviside to Lodge, 24 Sept. 1888, OJL-UCL; cf. Heaviside 1892, 1: ix–x and 2: 323n, and Nahin 1988: 148.

33. Heaviside to editor of *Phil. Mag.* [draft], 18 July 1887, OH-IEE.

34. Heaviside to Lodge, 24 Sept. 1888, OJL-UCL. The manuscript and "separate document" have apparently not survived.

35. Ibid. and the marginal annotations on a clipping of the *Nature* obituary of Preece, OH-IEE; see Nahin 1988: 287. In a letter of 4 Nov. 1950 to Lodge's former secretary Helen Alvey (OJL-UB), Heaviside's friend G. F. C. Searle said that no account of Heaviside's life could be at all adequate unless it recognized the great role animosity toward Preece played in his thinking.

36. Heaviside to editor of *Phil. Mag.* [draft], 18 July 1887, OH-IEE. Three appendixes by Oliver, which were not sent to Preece for clearance, were the only parts of the Heavisides' paper ever to appear in print; see Heaviside 1892, 2: 323–54.

iside, had to decline it. "However true it may be," he replied, "and however necessary, it is libellous," and the *Electrician*, already facing several lawsuits, could not risk another.[37] Heaviside then tried several other journals, but they too declined to publish his letter.[38] He also sent it to Silvanus Thompson, but while this tactic of appealing directly to scientists for support was later to work well for Heaviside, in this case it elicited little more from Thompson than expressions of horror at Preece's "impudence" and his "monstrous" behavior.[39]

Although he was prevented from taking Preece publicly to task for suppressing his "Bridge System" paper, Heaviside was still able to publish other papers on telephony, and he sought through these to expose Preece's scientific errors. He had been discussing induction balances in his regular series in the *Electrician*, but on 3 June 1887 he interrupted this because of what he called "the pressure of a load of material that has come back to me under rather curious circumstances."[40] This "load of material" was, of course, the paper Preece had blocked, and in his *Electrician* articles Heaviside began to give an account—the first actually to appear in print—of his theory of inductive loading and the distortionless circuit. He did not describe the "curious circumstances" under which this material had come back to him, but he made it clear that he had a low opinion of Preece—"the eminent scienticulist," as he called him—and his *KR* law approach to telephonic theory.[41]

Heaviside continued his articles in the *Electrician* through the summer, explaining in some detail how distortion occurs and how extra inductance could be used to reduce it. He also sought to publish similar but mathematically more sophisticated articles in the *Philosophical Magazine*, but without success. His series "On the Self-Induction of Wires" had been appearing there for nearly a year and had already run to much greater length than originally planned; when he sent an eighth installment in the summer of 1887 and mentioned plans for a ninth, it is perhaps not surprising that the editor declined it and suggested that such a long work might better be published as a book.[42] Heaviside wrote to

37. Biggs to Heaviside, 30 May 1887, OH-IEE; Biggs said he and other members of the *Electrician* staff already faced twelve libel suits.

38. Heaviside to editor of *Electrical Review* [draft], 3 June 1887, and editor of *Electrical Review* to Heaviside, 3 and 8 June 1887, OH-IEE; cf. Heaviside 1892, 1: x.

39. Heaviside to Thompson [draft], 7 June 1887, and Thompson to Heaviside, 9 and 13 June 1887, OH-IEE.

40. Heaviside 1892, 2: 119 [1887].

41. *Elec. 19* (1887): 79; when this was reprinted in Heaviside 1892, 2: 119, Heaviside changed "scienticulist" to "practician." "The eminent scienticulist" was one of Heaviside's many nicknames for Preece; others included "the Bouncer," "Mr. Prigs," and "Taffy."

42. William Francis [editor of *Phil. Mag.*] to Heaviside, 18 July 1887, OH-IEE. Part 8 of "On the Self Induction of Wires" was first published in Heaviside 1892, 2: 307–23.

explain the importance of distortionless propagation and to tell how Preece had prevented him from publishing on it in the *Journal of the Society of Telegraph-Engineers and Electricians;* would the *Philosophical Magazine* please reconsider its rejection? But he received no reply.[43]

The outlets available to Heaviside had thus dwindled to just one, the *Electrician.* Biggs continued to publish his articles every two weeks or so through the summer, but events in the fall of 1887 began to threaten this outlet as well and to put Heaviside in danger of being silenced altogether. At the September meeting of the British Association, Preece gave several papers in which he reiterated his adherence to the *KR* law and called again for the elimination of self-induction from telegraph and telephone lines. In the most substantial of these papers, "On the Coefficient of Self-Induction in Telegraph Wires," he claimed to have found experimentally that the inductance of copper wires was negligibly small and pointed to this as the reason for the good quality of signaling on them. "I have spoken with telephones over . . . 540 miles of copper wire," he said, "with a clearness of articulation that is entirely opposed to the idea of the presence of any measurable magnitude of *L*"; indeed, he said, it was the "virtual absence" of inductance that made copper wires so superior for telephony.[44]

This was more than Heaviside could bear. It was "dreadful stuff, rank quackery," he later told Lodge, and late in September he sent the *Electrician* a detailed and biting critique in which he declared that Preece was "wrong not merely in some points of detail, but radically wrong, generally speaking, in methods, reasoning, results, and conclusions."[45] After an introduction dripping with sarcastic praise for "William Henry Preece, F.R.S., the eminent electrician," Heaviside showed how Preece's lack of understanding of induction balances and other techniques for measuring self-induction had led him to assign to the inductance of copper wires a value that was several hundred times less than that derivable from quite elementary and well-recognized methods. Preece had trapped himself in a circle of fallacious reasoning: convinced that self-induction was prejudicial to clear signaling, he had taken the high quality of transmission over copper wires as proof that their inductance must be negligible and had then sought to bolster this conclusion by whatever means came to hand, including the very free manipulation of data. In fact, Heaviside said, Preece's basic premise was mistaken: signaling was actually *improved* by self-induction, and it was the relatively *high* ratio of

43. Heaviside to editor of *Phil. Mag.* [draft], 18 July 1887, OH-IEE; cf. Heaviside to Lodge, 24 Sept. 1888, OJL-UCL.
44. Preece 1887c; cf. Preece 1887a and 1887b.
45. Heaviside to Lodge, 21 Sept. 1888, OJL-UCL; Heaviside 1892, 2: 160–65.

inductance to resistance (L/R) in copper wires that made them work so well.[46] He closed with a table illustrating how even modest amounts of extra inductance could dramatically reduce telephonic distortion.

Biggs was reluctant to publish Heaviside's letter as first submitted; the sarcasm about Preece should be toned down, he said, and the controversy kept on a purely scientific plane.[47] But Biggs's advice, sensible as it was, soon became irrelevant. Early in October he was abruptly removed as editor of the *Electrician* and replaced by W. H. Snell, a young electrical engineer.[48] Biggs's support for Heaviside almost certainly contributed to his removal. A year later, as editor of the *Electrical Engineer*, he wrote that he had published Heaviside's articles in the *Electrician* "in spite of most strenuous opposition by the proprietors and every member of the staff" and hinted darkly at "interesting" circumstances surrounding his departure.[49] Whether Preece had interceded with John Pender and the other proprietors of the *Electrician* to have Biggs removed cannot now be determined, but he no doubt welcomed the subsequent disappearance from the journal of Heaviside's attacks on him.

Snell refused to publish Heaviside's critique of Preece's British Association paper, telling Heaviside on 22 October that although he agreed with him on "the purely scientific side of the question," it was "impossible to neglect other considerations in a matter of this kind."[50] Heaviside had no doubt that these "other considerations" involved Preece. When his letter was returned, he found, he said, that it had been in Preece's hands and that "the em. sci. [had] made his marks on it." Heaviside later told Lodge that he had "positive knowledge (from information received)" that Snell had "doctored communications, doctored them in Preece's favor."[51] Snell denied these allegations, telling Heaviside, "You are entirely mistaken in your surmises with regard to the relations between this journal and the individual in question [Preece]," and it is quite possible that Snell was not doing Preece's bidding but was simply concerned, as Biggs had been, about a possible libel suit.[52]

Unlike Biggs, however, Snell was not willing to continue publishing Heaviside's series "Electromagnetic Induction and Its Propagation,"

46. Ibid.; cf. Heaviside 1893–1912, 1: 442–44 [1893]. Iron wires were handicapped by the "skin effect," which greatly increased their resistance to high-frequency currents.
47. Biggs to Heaviside, 20 and 22 Sept. 1887, OH-IEE; the letter Biggs referred to was probably not quite the same as the one eventually published in Heaviside 1892, 2: 160–65, which is headed "Sept. 24, 1887; but now first published."
48. Printed announcements of the change of editors were sent to regular contributors; see Snell to Heaviside, 12 Oct. 1887, OH-IEE, quoted in Nahin 1988: 161–62.
49. [Biggs] 1889, quoted in Nahin 1988: 153. A notice in *Elec. 19* (1887): 469 stated that Biggs had resigned, but Biggs's later remarks make it clear that he was removed.
50. Snell to Heaviside, 22 and 27 Oct. 1887, OH-IEE.
51. Heaviside to Lodge, 21 and 24 Sept. 1888, OJL-UCL.
52. Snell to Heaviside, 27 Oct. 1887, OH-IEE.

which had been running in the *Electrician* since 1885. A regular installment was published on 7 October in the last issue edited by Biggs, but
when no more appeared in the next month and a half, Heaviside began
to worry. When he wrote to ask when he might expect to see the next
installment in print, Snell replied on 30 November that the series was
being canceled. Although he professed to rate "the intrinsic value of
your papers very highly indeed," Snell said he found that they attracted
very few readers; indeed, he told Heaviside, "I have not been able to
discover *any*."[53] The *Electrician* was a commercial journal with great
pressure on its available space, and Heaviside's highly mathematical and
idiosyncratic papers no longer had a place in it; indeed, it is perhaps
more surprising that Biggs had been willing to publish them than that
Snell was not. A brief final installment of "Electromagnetic Induction"
appeared on 30 December, concluding with a note stating, "The author
much regrets to be unable to continue these articles . . . having been
asked to discontinue them."[54]

This was the low point in Heaviside's struggle to be heard. After being
blocked at the Society of Telegraph-Engineers and Electricians and rejected by the *Philosophical Magazine,* he had now lost his main ally, Biggs
of the *Electrician,* and seen the series that was his life's work abruptly
terminated. He found himself "boycotted all round," he later said; he
lost the small income he had received from the *Electrician* and, even
worse, was left with no outlet for what he considered to be his most
important work. "I came, for a time," he said, "to a dead stop, exactly
when I came to making practical applications in detail of my theory, with
novel conclusions of considerable practical significance relating to long-
distance telephony."[55] Heaviside was sure that Preece was behind the
"peculiar concurrence and concatenation of circumstances" by which he
had been silenced, and he was intent on striking back. The blocking of
his work "was a serious matter to me last October," he told Lodge in
1888, and had he been unable to find an outlet for his writings, "then I
should have been obliged to take some very decisive measures," he said,
"and I am a determined character in my way."[56]

Campaigning for Recognition

Heaviside never said what "decisive measures" he had in mind in October 1887, and fortunately he never felt compelled to carry them out.

53. Heaviside to editor of *Elec.* [draft], 28 Nov. 1887, and Snell to Heaviside, 30 Nov.
1887, OH-IEE.
54. Heaviside 1892, 2: 155 [1887].
55. Ibid., 1: ix; cf. Heaviside to Lodge, 21 Sept. 1888, OJL-UCL.
56. Heaviside to Lodge, 24 Sept. 1888, OJL-UCL.

The *Philosophical Magazine* began to accept his work again in December, beginning with a paper "On Resistance and Conductance Operators" which included some remarks on Preece and distortionless propagation.[57] This was followed in 1888 by "Electromagnetic Waves," an important series of six papers that represented a deliberate and important change in Heaviside's publishing strategy.[58] "Up to last year," he told Lodge in 1888, "I never troubled myself in the least about credit, recognition, and so forth, but just tried to get as much of my work published as possible." But after the experience of "being sat upon" by Preece and having his "best work rejected," Heaviside concluded that "to prevent the possibility of a repetition of such a series of humiliations," he must acquire a prominent scientific reputation. "It is not enough," he said, "to be appreciated by a few who can judge, it seems; though intrinsically, I care for nothing else; . . . it seems to be necessary to have a reputation of another kind, at second hand, so to speak, from those who can to those who can't judge, and this is the explanation of my apparent sudden anxiety to have my superlative merits recognised!"[59] Heaviside's "Electromagnetic Waves" series was the first step in his deliberate campaign to attract the attention and support of other electrical scientists and so to move from the fringes of the community of telegraph engineers into the center of the community of electromagnetic theorists.

Heaviside decided that the best way to accomplish his aims was "to go to headquarters." If he could gain the attention of William Thomson, the acknowledged leader of British physics and electrical engineering, then the rest would follow; in particular, if he could win Thomson over to his theory of telephonic propagation, the opposition of Preece and his allies would be of relatively little moment. The fact that Thomson had praised Preece at the 1887 British Association meeting gave Heaviside an additional incentive to raise his sights and seek to convince Thomson himself that inductive loading could actually improve telephonic transmission. "Instead of shooting at a sparrow," he said, "I would try to bring down the Sun"; instead of tangling with Preece, he would go directly to Thomson.[60]

Heaviside's strategy worked. Late in November 1887 he sent the manuscript of the first two parts of "Electromagnetic Waves" to Thom-

57. Heaviside 1892, 2: 355–74 [1887].
58. Ibid., pp. 375–467 [1888].
59. Heaviside to Lodge, 21 Sept. 1888, OJL-UCL.
60. Ibid.; see *Elec. 19* (1887): 426 for an example of Thomson's praise for Preece at the BA meeting. Thomson had known since the late 1850s that inductance could affect transmission on telegraph lines, but he did not recognize its exact effects and benefits until after he read Heaviside's work in the late 1880s; see Smith and Wise 1989: 457–58 and William Thomson 1882–1911, 3: 487–88 [1889].

son, who was an advisory editor of the *Philosophical Magazine*. Thomson wrote back to ask Heaviside about some of his remarks about "the velocity of electricity" and Thomson's 1855 cable theory, and Heaviside "took the opportunity," he told Lodge, "and gave him my mind about it and his attitude toward modern electromagnetism, though not half so thoroughly as I wanted to because I was afraid of rejection."[61] Heaviside was convinced that Thomson was "on the wrong track" in electrical matters and that his failure to appreciate Maxwell's theory would, if it persisted, prove "disastrous for progress."[62] He said this as explicitly as he politely could in a postscript to the first installment of "Electromagnetic Waves," citing in support "a most moving appeal from Prof. G. F. Fitzgerald, in *Nature*, about three years since," in which FitzGerald had called on Thomson not to allow the weight of his authority to delay the acceptance of Maxwell's theory.[63] Although Heaviside had not yet met or even corresponded with FitzGerald, he already recognized him as a fellow Maxwellian and an ally in the struggle against what they both saw as Thomson's increasingly outmoded views.

Heaviside's fears of rejection proved unfounded, and his "Electromagnetic Waves" series, which ran throughout 1888, was in fact quite well received. Several physicists who were not readers of the *Electrician*, notably Rayleigh, first learned the extent of Heaviside's work on Maxwell's theory through these *Philosophical Magazine* articles.[64] In January 1889, Heaviside had fifty copies of the whole series printed and bound, together with another paper, "The General Solution of Maxwell's Electromagnetic Equations," and sent them out to a carefully selected list of British and foreign physicists.[65] That someone of Heaviside's very limited means would go to this expense underscores how serious he was about spreading an appreciation of his work among leading scientists. It was a "good stroke of business having the separate copies," he said later, and the demand soon exceeded the supply.[66]

By the time he began sending off copies of his "little book," Heaviside

61. Heaviside to Lodge, 21 Sept. 1888, OJL-UCL; see also Heaviside notebook 10, p. 135, OH-IEE.
62. Heaviside 1892, 2: 394 [1888].
63. Ibid.; see FitzGerald 1902: 170–73 [1885].
64. Lodge to Heaviside [draft], 23 Sept. 1888, OJL-UCL, reporting a remark of Rayleigh's.
65. Heaviside 1889; inscribed copies are held by University College London, the University of Washington, the FitzGerald Library of the Physics Department at Trinity College Dublin, and (according to O'Hara and Pricha 1987: 56) Karlsruhe University. Heaviside consulted Lodge about the mailing list (in Heaviside notebook E, OH-IEE); see Heaviside to Lodge, 19 and 23 Jan. 1889, OJL-UCL. Heaviside refused to send a copy to Preece: "I have immortalised him inside," he told Lodge, "and what more does he want?"
66. Heaviside notebook 3a, entry 161, OH-IEE; cf. Heaviside 1892, 1: v.

was already far more widely known and well regarded than when he had begun writing the articles in it just over a year before. In the intervening months, a series of unexpected experimental discoveries had moved electromagnetic waves from the realm of theoretical speculation into that of established fact and had made Heaviside's papers the focus of intense scientific interest. These discoveries had grown out of areas seemingly far removed from Heaviside's own work in telegraphy and telephony, and when Oliver Lodge announced the first of them in the spring of 1888, it must have seemed to Heaviside almost like a bolt from the blue.

Lightning

Lodge had been among the first to try to detect electromagnetic waves experimentally, but after the failure of his initial attempts to produce "electromagnetic light" in 1879 and 1880, he had let the effort drop for several years. Early in 1888 he was led back to it in a rather roundabout way by a request that he give two lectures on lightning protection. The Society of Arts in London had received a small bequest from the estate of Dr. Robert Mann, a South African who had been active in the erection of lightning rods, and it was thought that a pair of lectures by Lodge, whose skill as a scientific speaker was already well known, would make a fitting memorial. Lodge accepted the offer and began to prepare for the lectures by experimenting with Leyden jars, using their sparks to simulate lightning in order to test various methods of protection. It was later found that the resemblance between the spectacular flashes produced in this way and actual lightning bolts was more apparent than real, but considered simply as studies of electrical discharges, Lodge's experiments were of great scientific value.[67]

Lodge's most important experiments involved what he called "the alternative path."[68] He found that when a jar was given two or more paths along which to discharge, a sudden flash did not simply take the "easy" path offered by a thick copper rod; it rushed simultaneously along all available paths, even if that meant running along a high-resistance iron wire or sparking across an air gap. Lodge explained this surprising behavior by pointing to the oscillatory character of the discharge and the consequent influence of self-induction. William Thomson had shown theoretically in the 1850s that the discharge of a condenser did not sim-

67. Lodge 1931a: 96–97.
68. Lodge 1888c, repr. Lodge 1892a: 32–37.

ply flow in one direction but instead oscillated with a frequency determined by the capacitance, inductance, and resistance of the circuit. With each reversal, the current had to overcome its self-induction or "electromagnetic inertia"—in Lodge's view, the literal inertia of whirling machinery in the ether. In his Leyden jar discharges, the current reversed up to a million times per second, and Lodge pointed out that at such high frequencies, the ordinary frictional resistance of the wire was completely overshadowed by the inertial impedance caused by self-induction.

Conduction was further modified at high frequencies by the "skin effect," which had first come to light in 1886 in the wake of experiments by D. E. Hughes. While working with induction balances, Hughes had found evidence that rapidly varying currents were subject to higher resistance and lower self-induction than were steady currents; Rayleigh and Heaviside had used Maxwell's theory to show this effect resulted from a tendency of oscillatory currents to confine themselves to the outer surface of a conductor.[69] Heaviside viewed the skin effect as a straightforward consequence of his general theory of the action of the field and the function of wires. Since electromagnetic force and energy enter a conductor from the surrounding field, the flow of current begins along the boundary of the wire, Heaviside said, and then works its way inward. With rapid oscillations, self-induction prevents the current from penetrating beyond the outer skin of the wire before the driving force is reversed. The effect is strongest in iron because of its high magnetic "inertia"; at a frequency of 1,000 cycles per second, the current in an iron wire is almost entirely confined to a layer less than one-third of a millimeter thick.[70]

Lodge's discharge experiments provided especially strong evidence of the skin effect; at the frequencies he was working with, the current scarcely penetrated the outer surface at all. Under such conditions, the material composition of the conductor made relatively little difference, and an iron wire could become as good a path for a high-frequency discharge as a copper wire, whose resistance and inductance for steady currents are much lower. Lodge had begun his experiments firmly convinced that copper would prove far superior to iron for lightning conductors, and he was astonished to find that it did not.[71] He said later that

69. See Jordan 1982b.
70. Heaviside 1892, 1: 440 [1885]; cf. Feynman 1964, 2: 32.11.
71. Lodge 1888c, repr. Lodge 1892a: 36, 40–41. In an early installment of his "Modern Views of Electricity," *Nature 37* (1887): 10, Lodge had declared that the high inductance of iron made it "about 90,000 times worse than copper for the purposes of a lightning conductor," but when this section was reprinted, he replaced this passage with a brief account of the skin effect; see Lodge 1889b: 101–2.

his main aim in undertaking the Mann lectures had been to call attention to the role of self-induction in lightning protection, and his experiments convinced him that it was of even more importance than he had first imagined.[72]

Lodge's first lecture on 10 March 1888 was mainly historical and descriptive, but in the second, delivered a week later, he broke new ground and argued forcefully that self-induction ought to be a primary consideration in the design of lightning conductors. Punctuating his presentation with dramatic Leyden jar flashes, he attacked many of the orthodox rules of lightning protection as at best a waste of money and at worst a positive danger. The fat copper rod usually recommended might be an excellent conductor for steady currents, but Lodge asserted that the rapid oscillations of a lightning bolt were handled better by a cheap iron wire. Neglect of self-induction had led, he said, to practices that could result in side sparking and fires and which in some circumstances were more dangerous than having no conductor at all.[73] He raised grave doubts about the adequacy of the protection existing conductors were providing and called for a thorough overhaul of the standard rules governing their design and installation.

Such statements naturally aroused controversy, and Lodge soon found himself in a protracted battle with defenders of the traditional methods of lightning protection. Preece emerged as his most active adversary; he had taken a leading part in the 1881 Lightning Rod Conference at which the standard rules for protection were formulated, and in the debate with Lodge, as elsewhere, he was an outspoken opponent of what he called the Maxwellians' "mania . . . for self-induction."[74] Many of Preece's arguments were fallacious, but it should be noted that the orthodox rules he defended were eventually found to be preferable in many respects to those proposed by Lodge. While Leyden jar discharges indeed oscillate just as Lodge said they did, lightning bolts do not, and in going from the laboratory to a thunderstorm, Lodge had made a serious misstep.[75] It was many years before these issues were settled, however, and in the meantime the debate over lightning conductors stimulated some very valuable work, not least the experiments that led Lodge to the discovery of electromagnetic waves on wires.

Lodge first saw signs of waves on wires in February 1888 while trying a

72. Lodge, in Preece et al. 1888: 647.
73. Lodge 1892a: 65 [1888]; in the "Preface," Lodge denied having ever said that older designs were "useless or dangerous," but that is the clear implication of several of his statements in the Mann lectures.
74. Preece, in Preece et al. 1888: 646.
75. See Lodge 1931a: 96–97 and Jordan 1982a: 453–54.

variant of his alternative-path experiment. He noticed that the discharge of a Leyden jar through a short wire caused large sparks to jump a gap between the free ends of wires attached to the jar and extending some distance away from it; moreover, the sparks could be made weaker or stronger by adjusting the length of the wires, suggesting that resonance was involved. Prodded by his junior colleague A. P. Chattock, also an ardent Maxwellian, Lodge realized that the discharges were emitting electromagnetic waves into the ether, just as FitzGerald had predicted theoretically in 1883, and that these waves were surging along through the space around the wires. Lodge had, he later said, "hit upon an arrangement which, without any thought or scheming at all, gave me evidence of the very waves I had been thinking so much about, and enabled me to measure their lengths, though not in a previously planned-out way."[76] The sparking at the free ends was due to a "recoil kick" as the waves were reflected back, and Lodge showed at his second Mann lecture that if the length of the wires and the oscillation time of the discharge were properly adjusted, a pattern of standing waves could be established and the sparking made quite dramatic. He later found that, with long wires and a strong discharge, the space around the wires actually glowed, so that in a darkened room, one could pick out the nodes and antinodes as clearly as if they were waves on a string.[77]

In following up his wave experiments and exploring their relation to the skin effect, Lodge began to look more deeply into the theory of propagation along wires. He happened to see the first installment of Heaviside's "Electromagnetic Waves" in the February 1888 *Philosophical Magazine* and was led to read for the first time some of Heaviside's other papers. Lodge was impressed by what he found, and in his second lecture he went out of his way to remark on "what a singular insight into the intricacies of the subject, and what a masterly grasp of a most difficult theory, are to be found among the eccentric, and in some respects repellant, writings of Mr. Oliver Heaviside."[78] This was one of the first times public attention had been drawn to Heaviside's work, and despite the reservations about his expository style, it gave his campaign for recognition a welcome boost. The preceding year had been a dark time for Heaviside, and he later told Lodge, "I looked upon your 2nd lecture when I read it as a sort of special Providence!"[79]

Heaviside saw only fragmentary reports of Lodge's lectures at the

76. Lodge 1888a.

77. Lodge 1892a: 59–62 [1888]; Lodge 1931b: 183–84.

78. Lodge 1888c: 236. When this passage was reprinted in Lodge 1892a: 47, Lodge deleted the words from "eccentric" through "repellant."

79. Heaviside to Lodge, 24 Sept. 1888, OJL-UCL.

time they were delivered, and early in June, in one of his first letters to Lodge, he wrote to ask where he might find a full and accurate text.[80] The lectures were finally published two weeks later, and when Lodge sent him a copy, Heaviside was pleased to see how fully Lodge's experiments confirmed his own ideas about the propagation of waves on wires. He wrote to congratulate Lodge for having made "a very substantial addition to our knowledge of sudden discharges" and to explain some points about self-induction and the skin effect. In particular, he pressed his claim for priority as the first to have predicted and explained the skin effect, which he had described in connection with his theory of the rise of current as early as January 1885, long before Hughes's experiments. He had not put much stress on it at the time, he told Lodge, because "I thought much more of my other discovery, the transfer of energy, which Prof. Poynting is to be congratulated upon, as I later found," but Heaviside now hoped that acknowledgment of his priority on the skin effect would draw attention to his other work and aid his campaign for recognition.[81]

Lodge was sympathetic to Heaviside's priority claims, and while he said to was "no part of my business to play the historian of science," he tentatively endorsed them in a letter to the *Electrician* early in July.[82] He also repeated and amplified his praise for Heaviside's command of electromagnetic theory, asserting that "no person now alive . . . has anything like the intimate familiarity" with the phenomena of alternating currents that Heaviside possessed. Remarking that "the experimental genius of Prof. Hughes is a matter of notoriety," Lodge declared that "the mathematical genius of Mr. Heaviside is equally striking, and will shortly, I believe, be as cordially recognised."

As they exchanged compliments in the summer of 1888, Lodge and Heaviside began to develop a relationship of mutual support and dependence. At a time when Heaviside was almost desperate for recognition, Lodge began to provide it dramatically and effectively through his lectures and publications. Heaviside was grateful and even began to hope that the burst of interest in his work might prompt the *Electrician* to renew his canceled series on electromagnetic propagation. "I was in-

80. Ibid., 5 June; the only earlier surviving letter from Heaviside to Lodge is one of 15 Jan. 1885, OJL-UCL, thanking Lodge for a copy of a paper on contact potential. Lodge's Mann lectures were not released for publication until 22 June 1888; see *Elec. 21* (1888): 197.

81. Heaviside to Lodge, 27 June 1888, OJL-UCL; cf. his first mention of the skin effect in Heaviside 1892, 1: 440 [1885].

82. Lodge 1888d; but see also Lodge 1888b: 217n, which lists several others who could claim to have predicted the skin effect, though Lodge acknowledged that Heaviside's 1885 statement was the clearest and fullest.

formed substantially that no one read my articles," he noted dryly in a letter to the editor in August. "Possibly some few may do so now, with Dr. Lodge's experiments in practical illustration of some of the matters considered."[83] In return, Heaviside provided Lodge with the theoretical underpinnings he needed to support his controversial claims about the behavior of oscillating currents. Lodge came to rely on Heaviside for the theoretical interpretation of his own experiments, and in his writings on electrical discharges over the course of 1888 and 1889, he drew more and more heavily on Heaviside's ideas about energy flux and the function of wires.[84] A clear division of labor was developing within the Maxwellian group, with Lodge taking the role of chief experimenter and publicist and Heaviside that of chief theoretician.

Lodge continued his Leyden jar experiments after the Mann lectures and in July sent the *Philosophical Magazine* a paper "On the Theory of Lightning Conductors," in which he summarized his main findings. Citing Heaviside for the general theory of electromagnetic propagation, he described in more detail his "alternative path" and "recoil kick" experiments and stated explicitly that condenser discharges "disturb the surrounding medium and send out radiations, of the precise nature of light," though with wavelengths of several meters.[85] It might be possible, he said, to detect such waves in free space if they could be properly concentrated; in any case, he planned to continue to experiment with waves on wires and promised to give a full account of his measurements of their wavelengths at the Bath meeting of the British Association in September. Lodge believed he was on the track of a major discovery, and as he left for a holiday in the Alps in mid-July, he was confident that his wave experiments would be the hit of the upcoming meeting. But the electromagnetic waves that caused a sensation at Bath were, as it turned out, not Lodge's but those discovered by a young German experimenter, Heinrich Hertz.

83. Heaviside 1892, 2: 486–88 [1888].
84. Lodge 1892a: 111, 219 [1888–89].
85. Lodge 1888c: 275–76; Lodge 1888b.

CHAPTER SEVEN

Bath, 1888

The Bath meeting of the British Association was in many ways the climax of the Maxwellian story. Several seemingly disparate chains of events came together there early in September 1888 and, within the space of a week, combined to vault Maxwell's theory and its proponents into a position of unaccustomed prominence in the scientific world. FitzGerald opened the physics sessions in dramatic fashion on 6 September by announcing Heinrich Hertz's "epoch-making" discovery of electromagnetic waves. Lodge followed the next day with an account of his own experiments on waves along wires and on the effect of self-induction on oscillating currents, and he later faced Preece in a spirited debate that ostensibly concerned lightning conductors but in fact turned on the much deeper issue of "practice" versus "theory." And throughout the meeting, William Thomson, FitzGerald, Lodge, and Henry Rowland wrestled with one of the most difficult points in Maxwell's theory, that of the propagation of electric potential—debating, as FitzGerald put it, whether they ought to "murder ψ." The 1888 meeting was not the largest or most spectacular in the history of the British Association, but for the Maxwellians, it was, as Lodge said, "one of the most interesting and important that have ever been held."[1]

Although Heaviside was not at Bath (he never attended public meetings), his name came up repeatedly and he was an important offstage presence. His writings were just beginning to attract public notice in the summer of 1888, and the events at Bath greatly accelerated his emer-

---

1. Lodge 1888e: 663.

gence, along with FitzGerald and Lodge, as one of the acknowledged authorities on Maxwell's theory and its application to electromagnetic waves. The experiments of Hertz and Lodge were, Heaviside later said, "the means of stirring up an amount of interest in this theory that was quite wonderful to witness," particularly for one who, like himself, had been pursuing the theory for years without stirring up much interest at all.[2] In the months after the Bath meeting, Maxwell's theory—or more properly, the revised "Maxwellian" version of it—became far more widely known and accepted than ever before, while the Maxwellian group itself became both more cohesive and more influential.

Hertz's Waves

When Lodge completed his paper "On the Theory of Lightning Conductors" early in July 1888 and prepared to leave for a few weeks of hiking in the Alps, he was confident that his experiments on electromagnetic waves on wires were new and important and that they would be hailed as great discoveries at the upcoming meeting of the British Association. But he soon found that he had been anticipated. "When going away from Liverpool on a holiday," he wrote later, "I read in the train the July number of Wiedemann's *Annalen*, and there found that Dr. Hertz had obtained much better and more striking evidence of these electromagnetic waves" and had displayed them not only along wires but in free space as well.[3] Lodge's own experiments were rendered "superfluous," he said, except as illustrations of some relatively minor points, and his hopes for acclaim at Bath were largely snuffed out. But his disappointment at finding that he had been beaten to the prize he had so long pursued was tempered by his pleasure at the beauty of Hertz's experiments. On 24 July, while still at Cortina, Lodge sent the *Philosophical Magazine* a brief note in which he called attention to Hertz's paper and to the impressive discoveries it contained. "The whole subject of electrical radiation," he declared, "seems working itself out splendidly."[4]

Lodge and his British colleagues were surprised to find the confirmation of Maxwell's theory they had so long sought appearing first in the pages of a German journal. Germany had been a stronghold of action-at-a-distance theories for decades, and the field ideas of Faraday and Maxwell had made little headway there up to the time Hertz began his experiments. Heaviside regarded Germany as the place "where one

2. Heaviside 1892, 1: v.
3. Lodge 1888a.
4. Lodge 1888b: 230.

would least expect such a discovery to be made, if one judged only by the old German electro-dynamic theories," while Lodge remarked, in 1894, "Even now it is surprising that a continental philosopher should have reaped the fruits of the theoretical seed sown in England."[5] But the brilliance of Hertz's work was undeniable, and the British Maxwellians expressed only the highest admiration for it. They may have been disappointed that electromagnetic waves were not first detected by one of their own countrymen, but they could console themselves with the knowledge that Hertz was, as Lodge said, "no ordinary German."[6]

Heinrich Hertz was born in Hamburg in 1857, the son of a prominent jurist, and received an excellent and well-rounded early education.[7] After originally preparing to become an engineer, he switched to physics in 1877 and, after a year at Munich, went to Berlin to study under Hermann von Helmholtz. He showed great talent in both theory and experiment and was soon looked upon as one of the most promising of the younger German physicists. Lodge, who met him on a tour of Germany in 1881, was impressed by Hertz's grasp of physics and also by his friendliness and modesty.[8]

Hertz spent nearly six years at Berlin, and it was under Helmholtz's guidance that he began the work that eventually led him to the discovery of electromagnetic waves. Like most Continental physicists, Hertz was introduced to Maxwell's theory through the generalized system of electrodynamics Helmholtz had devised in 1870, in which each of the main competing electrical theories was treated as a special case corresponding to a particular value of the parameter "k": Wilhelm Weber's theory for $k = -1$, Franz Neumann's for $k = 1$, and Maxwell's for $k = 0$.[9] Helmholtz's theory was based on action-at-a-distance principles, and the resemblance between one of its special cases and Maxwell's field theory was in fact more apparent than real. No one fully realized this until after 1888, however, and in the meantime Helmholtz's approach was accepted on the Continent as a convenient way to assimilate the unfamiliar British ideas within existing theoretical patterns.

    5. Heaviside 1893–1912, 1: 5 [1891]; Lodge 1894.
    6. Lodge 1889a: 532. When this was reprinted in Lodge 1889b: 368, Ludwig Boltzmann and other "ordinary Germans" (Boltzmann was in fact Austrian) took exception to the phrase "no ordinary German"; see Lodge's letter of apology, "International Courtesy," *Nature 50* (1894): 399, in which he denied any "rancour was felt in this country that the fruits of Maxwell's theory should have been reaped by a German." The offending phrase was deleted from the 2d ed.; see Lodge 1892b: 426.
    7. See Hertz 1977: 1–21.
    8. Lodge 1931b: 154.
    9. Hertz 1893: 20–28; Buchwald 1985a: 177–86. In the "Maxwell case," it is also necessary to make the polarizability χ_0 infinite.

HEINRICH HERTZ, 1891. From Deutsches Museum, courtesy AIP
Niels Bohr Library

"Dr. Hertz . . . has immortalized himself by a brilliant series of investigations
which have cut right into the ripe corn of scientific opinion in these islands."
—Oliver Lodge, Royal Institution lecture, March 1889

The most important immediate effect of Helmholtz's work was to sharpen the conflict among the three competing theories of electrodynamics. Thus it became an urgent matter in the 1870s for German physicists to find a way to distinguish experimentally between the case corresponding to Maxwell's theory and those corresponding to the theories of Weber and Neumann—far more urgent than in Britain, where the action-at-a-distance theories were by then already regarded as obsolete. All three theories yielded almost identical observable consequences, one of the few distinguishing features of Maxwell's theory being its prediction that dielectric displacement currents should give rise to electromagnetic effects. In 1879, Helmholtz urged Hertz to look for such effects, but Hertz (then only twenty-two) demurred; for after a close study of the theory and of the available experimental techniques, particularly those involving high-frequency oscillations, he had concluded that the effect would be too small to detect.[10] Hertz kept the problem in mind, however, and remained alert for any new methods that might enable him to solve it.

In the early 1880s, particularly during two years he spent as a *Privatdozent* at Kiel, Hertz had few facilities for electrical experimentation and instead concentrated on purely theoretical studies, including an important 1884 paper on Maxwell's theory—in its Helmholtzian version, of course.[11] In 1885 he moved to the Technische Hochschule in Karlsruhe, which had an excellent laboratory, and there he gradually returned to experimental work. Late in 1886 he made an important observation: while examining some of the apparatus used in lecture demonstrations, he noticed that the oscillatory discharge of a Leyden jar or induction coil through a certain wire loop caused sparks to jump a gap in a similar loop a short distance away. He recognized this as a resonance phenomenon and saw that such sparking loops could serve as very sensitive detectors of oscillating currents and even, as he later found, of electromagnetic waves. They provided what FitzGerald had sought without success in 1884: "something to *feel* these rapidly alternating currents with."[12]

Hertz's observation was in many ways a matter of luck; none of the British proponents of Maxwell's theory had guessed that such high-frequency oscillations would be strong enough to produce visible sparks, and Hertz himself later said that his discovery of electromagnetic waves had rested in part upon "a special and surprising property of the electric

10. Hertz 1893: 3.
11. Hertz 1896: 273–90 [1884]; cf. D'Agostino 1975: 286–96 and Jungnickel and Mc-Cormmach 1986, 2: 44–48.
12. FitzGerald to J. J. Thomson, 23 Dec. 1884, quoted in Rayleigh 1943: 22.

spark which could not be foreseen by any theory."[13] Few, however, were as well equipped as Hertz to follow up this chance observation and to turn these tiny sparks into such a powerful experimental tool.

Hertz's spark loops provided just what he needed to answer the question Helmholtz had posed in 1879, and by the end of 1887 he had succeeded in demonstrating that Maxwell's dielectric displacement currents indeed produced electromagnetic effects. Gradually, however, he came to see that "the gist and special significance" of Faraday's and Maxwell's field theory lay not in ascribing electromagnetic actions to material dielectrics but in ascribing them also to empty space, or rather to the ether.[14] The real test of Maxwell's theory, Hertz realized, would be to produce and detect electromagnetic waves in free space. His oscillator and detector loops were easily adapted to this purpose, and early in 1888 he published an account of experiments in which he demonstrated interference between waves traveling along a wire (Hertz said "*in* the wire," indicating that he did not yet have a thoroughly Maxwellian view of the phenomenon) and those transmitted through the air.[15] He was able to show that electromagnetic forces propagated at a finite velocity, but there were difficulties in the interpretation of these experiments; the waves seemed to travel much more slowly along the wire than through the air, making the interference complicated and suggesting that electrostatic and electromagnetic forces propagated at different velocities. Hertz was troubled by this discrepancy, and only much later was he able to trace it to the distorting effects of an iron stove that sat near his wire.[16]

No such complications marred his next set of experiments, completed in March 1888 and published in July in the *Annalen der Physik* as "Ueber elektrodynamische Wellen in Luftraume und deren Reflexion" ("On Electromagnetic Waves in Air and Their Reflection"). Hertz now dispensed with the wire and simply reflected the waves off the walls of his lecture hall, setting up a pattern of standing waves whose nodes and antinodes he could locate with his loop detector. He found the wavelength to be about 9 meters, in good accordance with theoretical expectations. These experiments were fully analogous to well known interference experiments with sound, light, and waves on strings, and they allowed Hertz to display the propagation of electromagnetic forces in what he called "a visible and almost tangible form."[17] The simplicity of

13. Hertz 1893: 3; cf. FitzGerald 1902: 237 [1888].
14. Hertz 1893: 7.
15. Hertz 1893: 107–23 [1888].
16. Hertz 1893: 9.
17. Hertz 1893: 124 [1888].

these experiments made them especially impressive, and when in succeeding months Hertz found ways to show that electromagnetic waves
could be reflected, refracted, diffracted, and polarized exactly like the
much shorter waves that constitute light, his confirmation of Maxwell's
theory seemed complete.

Hertz was encouraged in his work by Helmholtz, but few others in
Germany in the 1880s were equipped to see the real significance of his
experiments; indeed, Hertz himself did not grasp it fully at first. Without clear guidance, it was difficult for most German physicists to draw
the connection between Hertz's initial experiments and the unfamiliar
doctrines of Maxwell's theory. The remark often made later, that "news
of Hertz's discoveries . . . reached Germany by way of England," was no
doubt an exaggeration, but it contained a large element of truth.[18]
FitzGerald, Lodge, and Heaviside had been studying the theory of electromagnetic waves for years and had achieved a remarkably clear and
detailed understanding of their properties by 1888; they were more
than ready to respond to any sign of their experimental detection.
Hertz's work "cut right into the ripe corn of scientific opinion in these
islands," Lodge declared in 1889; and well before German physicists had
taken much notice of Hertz's experiments, the British Maxwellians were
hailing them as a major breakthrough.[19] "Had it not been for the pre-
paredness of British physicists in this matter," Lodge later observed,
"Hertz's discovery might have been left to leak out gradually, and to be
more or less misunderstood for the course of ten or twenty years, as
usual."[20] Instead, the British Maxwellians moved quickly to give Hertz's
experiments the widest possible publicity and to label them from the
first as a decisive new confirmation of Maxwell's theory.

Reception

When Lodge read the July issue of the *Annalen der Physik* on the train
out of Liverpool, he was not in fact the first British physicist to learn of
Hertz's work and to appreciate its significance. A little over a month
before, FitzGerald had seen a brief mention of Hertz's earlier paper,
"Ueber die Ausbreitungsgeschwindigkeit der elektrodynamische Wir-
kungen" ("On the Finite Velocity of Propagation of Electromagnetic Actions"), and had written to ask for a copy of this "splendid verification of
Maxwell's theory." FitzGerald praised the work lavishly even before he

18. D. E. J[ones], "Heinrich Hertz" [obituary], *Nature 49* (1894): 265–66.
19. Lodge 1889a, repr. Lodge 1889b: 368.
20. Lodge 1894.

had seen the full version. "I consider that no more important experiment has been made this century," he told Hertz. "Your experiment will be called 'Hertz's classical experiment that decided between theories of electromagnetic action at a distance and by means of the ether.' "[21] This enthusiasm was perhaps premature; the experiments with waves on wires which Hertz described in the paper were in fact rather ambiguous and hardly qualified as a "splendid verification" of Maxwell. But in his reply, Hertz called attention to his more recent experiments with waves in free space, and these indeed provided the hoped-for confirmation of Maxwell's theory and of FitzGerald's own 1882–83 work on electromagnetic radiation from condenser discharges.[22]

FitzGerald was eager to draw attention to the importance of Hertz's discoveries, and he was soon given an excellent and unexpected opportunity to do so. Arthur Schuster had been elected president of Section A (Mathematics and Physics) of the British Association for 1888 and was expected to devote his opening address at the September meeting to an account of his recent work on electrical conduction through gases. He fell ill during the summer, however, and in what proved to be a very fortunate choice, FitzGerald was named to replace him.[23] No one was better equipped than FitzGerald to bring the news of Hertz's experiments before the scientific world, and the British Association provided an ideal forum for such an exercise in scientific publicity.

For more than fifty years, the annual meetings of the British Association had served as both serious professional conferences and traveling popular science shows. The 1888 meeting in the old resort city of Bath was a little smaller than most, with a total attendance of just under two thousand, but it was lively, interesting, and very fully covered in the national press. All agreed that "the greatest sensation of the meeting" was the display of Edison's improved wax phonograph and the competing "graphophone" of C. S. Tainter; hundreds, *Nature* said, flocked "to see and hear the wonderful little cylinders."[24] Several African explorers drew crowds to Section E (Geography), while the sessions of Section F (Economics) were enlivened by "The Transition to Social Democracy," a paper by the young George Bernard Shaw. But the real center of scientific interest at the meeting was electricity, above all, electromagnetic

21. FitzGerald to Hertz, 8 June 1888, HH-DM; cf. Hertz 1893: 107–23 [1888]. Hertz's paper had been presented by Helmholtz to the Berlin Academy of Sciences on 2 Feb. 1888; an offprint of it with FitzGerald's annotations is in FG-TCD Physics.

22. Hertz to FitzGerald, 11 June 1888, FG-RDS.

23. FitzGerald 1902: 229 [1888]. Hertz thought from FitzGerald's remarks that Schuster, an old friend of Hertz's, had died; see Hertz to FitzGerald, 23 Sept. 1888, FG-TCD Library.

24. *Nature 38* (1888): 469.

waves. Advances in electrical technology had helped make the phenomena of alternating currents and electromagnetic propagation the focus of attention in both the physics and engineering sections, and the news of Hertz's discovery fell on especially fertile ground at Bath.

When FitzGerald stepped before the assembled physicists of Section A on the morning of 6 September, he had two main aims: to publicize Hertz's experiments, then almost unknown even to specialists, and to bring out their significance for theories about the nature of electromagnetic forces. Maxwell had firmly believed that such forces were exerted through an intervening medium rather than directly at a distance, but he had been unable to point to any decisive evidence to support his belief. Now such evidence had been found, and FitzGerald announced, "The year 1888 will be ever memorable as the year in which this great question has been experimentally decided by Hertz in Germany, and, I hope, by others in England."[25] "By a beautiful device," he said, Hertz had generated and detected waves only a few meters long and, by demonstrating their interference, had established that they traveled at the speed of light. This finite speed of propagation was the key point, FitzGerald said; for it meant that the energy of the waves must reside in the intervening field and so proved that there must be a medium within which electromagnetic forces subsist. "It is a splendid result," he declared: "Henceforth I hope no learner will fail to be impressed with the theory—hypothesis no longer—that electromagnetic actions are due to a medium pervading all known space, and that it is the same medium as the one by which light is propagated; that non-conductors can, and probably always do, as Professor Poynting has taught us, transmit electromagnetic energy."[26] Hertz's experiments refuted once and for all the idea of electromagnetic action at a distance, FitzGerald said, and put the focus of further inquiry firmly on the ether. They constituted a striking confirmation of Maxwell's theory and a singular triumph for the Maxwellian program.

William Thomson, Rayleigh, and other leading physicists joined in praising Hertz's work at the meeting, and the chorus of acclaim was soon echoed in the press. *Nature* declared that FitzGerald had rendered "the greatest possible service to electrical science" by drawing attention to Hertz's experiments, and *Science*, the *Electrician*, the *Athenaeum*, and even the *Times* joined in hailing as "epoch-making" the discoveries announced at Bath.[27] Within a week after FitzGerald's address, and in direct re-

25. FitzGerald 1902: 231 [1888].
26. Ibid., p. 237.
27. *Nature 38* (1888): 577; *Science 12* (1888): 163–64; *Elec. 21*(1888): 548; *Athenaeum*, 15 Sept. 1888, p. 356; *Times*, 14 Sept. 1888, p. 6.

sponse to it, the *Electrician* began to publish an extended account of Hertz's experiments by G. W. de Tunzelmann, and translations of Hertz's papers soon began to appear in other British journals.[28] The Maxwellians took the lead in promoting the dissemination of Hertz's papers; Lodge translated one for *Nature*, and his assistant J. L. Howard another for the *Philosophical Magazine*, while it was at FitzGerald's urging that Hertz's important paper "Ueber Strahlen elektrischer Kraft" ("On Rays of Electric Force") was translated for the latter journal early in 1889.[29] The collection of Hertz's papers that appeared in Germany in 1892 was quickly translated into English and, graced with a preface by Lord Kelvin (as William Thomson had by then become) and retitled *Electric Waves*, it was well received and widely read.[30]

British physicists were not content just to read about Hertz's experiments, however, and soon set about repeating and extending them. FitzGerald and his assistant Frederick Trouton were especially active in this work, and by early 1889 they had not only repeated most of Hertz's main experiments but even displayed some of the more dramatic ones publicly at Trinity College Dublin and the Royal Dublin Society.[31] Lodge and his group at Liverpool experimented with new detectors and used huge pitch lenses to concentrate the electromagnetic radiation, while J. J. Thomson got an enthusiastic response when he demonstrated the waves to his lecture classes at Cambridge.[32] Lodge may have been exaggerating when he said that Hertz's experiments could be performed by "anyone with an induction coil, an empty room, and some bits of metal," but they were certainly widely repeated in Britain in 1888 and 1889.[33]

Hertz had expected that his experiments might eventually attract attention in Germany, where they ran against the prevailing tide of action-at-a-distance theories, but he was gratified and somewhat surprised by the enthusiastic response his work received in Britain. "I feared," he told

28. De Tunzelmann 1888.

29. Hertz, "The Forces of Electric Oscillations Treated according to Maxwell's Theory," trans. O. J. Lodge, *Nature* 39 (1889): 402–4, 450–52, 547–48; Hertz, "On the Propagation of Electric Waves through Wires," trans. J. L. Howard, *Phil. Mag. 28* (1889): 117–27; Hertz, "On Rays of Electric Force," *Phil. Mag. 27* (1889): 289–98; FitzGerald to Hertz, 14 Jan. 1889, HH-DM.

30. Hertz 1892. The translation by D. E. Jones appeared late in 1893; Kelvin suggested the title *Electric Waves*; see O'Hara and Pricha 1987: 103–8.

31. *Nature 39* (1889): 349–50; F. T. Trouton, "Repetition of Hertz's Experiments, and Determination of the Direction of the Vibration of Light," *Nature 39* (1889): 391–93; FitzGerald to Hertz, 14 and 23 Jan., 8 Feb. 1889, HH-DM.

32. E. J. Dragoumis, "Note on the Use of Geissler's Tubes for Detecting Electrical Oscillations," *Nature 39* (1889): 548–49; O. J. Lodge and J. L. Howard, "On Electric Radiation and Its Concentration by Lenses," *Phil. Mag. 28* (1889): 48–65; Rayleigh 1943: 43.

33. Lodge 1890b; see also Rowlands 1990: 5–8.

FitzGerald shortly after the Bath meeting, "that people there would say: these things are well known" and would regard the actual detection of electromagnetic waves as superfluous.[34] But it was in fact precisely those British physicists who already took the existence of electromagnetic waves for granted who were the most alert to the real value of Hertz's experiments. The Maxwellians knew how narrow a base of observational evidence they possessed before 1888, and they were sensitive to the criticism, cited by Heaviside, that they were "working out a mere paper theory."[35] Hertz's experiments provided a convincing answer to this reproach; indeed, Lodge declared that they constituted "a sort of apotheosis of Maxwell's theory," and the Maxwellians sought to ensure that they were seen in that light.[36] The discovery of electromagnetic waves greatly enhanced and broadened the prestige of the entire Maxwellian program, and the emergence of Maxwell's theory as *the* theory of electromagnetism, and of FitzGerald, Lodge, and Heaviside as its leading interpreters, dates from 1888.

"The Murder of ψ"

Besides broadly confirming Maxwell's theory, Hertz's experiments helped focus attention on several specific theoretical questions, in particular those associated with electromagnetic propagation. The Maxwellians had been pondering propagation questions for years, and FitzGerald, Poynting, and Heaviside were convinced even before 1888 that the treatment Maxwell had given in his *Treatise* was defective in important respects. They did not yet fully agree, however, on exactly how it should be modified—in particular, on whether the electrostatic potential ψ should be dropped from the propagation equations. The discovery of electromagnetic waves helped bring this rather arcane issue to the surface and touched off a lively and influential debate at Bath over what FitzGerald dubbed the "murder of ψ." This debate not only clarified an important point in electromagnetic theory but also played a key role in establishing the Maxwellians as the acknowledged authorities on such questions.

The reexamination of electromagnetic propagation began at Bath on 7 September when William Thomson presented "A Simple Hypothesis for Electro-Magnetic Induction of Incomplete Circuits." Although he

34. Hertz to FitzGerald, 23 Sept. 1888, FG-TCD Library.
35. Heaviside 1893–1912, 1: 6 [1891].
36. Lodge, introduction to his trans. of Hertz, "The Forces of Electric Oscillations Treated according to Maxwell's Theory," *Nature 39* (1889): 402.

had laid much of the groundwork on which Maxwell's theory was built, Thomson long remained skeptical of some of Maxwell's leading ideas. In his 1884 Baltimore lectures he had dismissed "the so-called electromagnetic theory of light" as "rather a backward step," and the notion of displacement currents was, he said, simply too speculative to be legitimate.[37] When he subjected Maxwell's theory to especially close scrutiny in the summer of 1888, his doubts about displacement currents only grew stronger. He responded at Bath with a "hypothesis" that indeed seemed simple: Maxwell's "curious and ingenious but ... not wholly tenable hypothesis" of displacement currents should be abandoned and only conduction currents be admitted to exist.[38] As Lodge put it in the "Sketch" of the meeting he wrote for the *Electrician*, Thomson, "finding ... certain apparent excrescences and redundancies" in Maxwell's theory, had sought "to lop them off and keep only that which is essential."[39] But as Poynting observed a short time later, Thomson's hypothesis would in fact "lop off not only Maxwell's excrescences but his whole theory"; without the keystone of displacement currents, the whole structure would fall to the ground.[40]

Thomson's "Simple Hypothesis" was certainly "very anti-Maxwellian," as Lodge said shortly after the meeting, but there was reason to think it represented "only a transition stage through which he was somewhat rapidly passing, and through which he may now have almost passed."[41] Thomson had in fact begun to retreat from the paper even before he delivered it. He later explained that he had originally intended to characterize Maxwell's hypothesis of displacement currents as "wholly untenable." But after arriving at Bath, "two days of the British Association before my paper was read gave me the inestimable benefit of conversation with others occupied with the same subject, and of hearing Professor Fitzgerald's presidential address in Section A, by which I was helped happily to modify my opinion"; in reading his paper to the section, he softened his words to "not wholly tenable."[42] *Nature* reported that this "very modified admission ... that, after all, Clerk Maxwell may

37. William Thomson, "Lecture I" [1884], in Kargon and Achinstein 1987: 12.

38. William Thomson 1882–1911, 4: 539–44 [1888], on p. 543. See Smith and Wise 1989: 474–82 for an account of Thomson's analysis of Maxwell's theory in the summer of 1888.

39. Lodge 1888e: 624.

40. Poynting to Lodge, 12 Oct. 1888, in Poynting 1920: 246 [1888].

41. Poynting 1920: 248 [1888], quoting a letter from Lodge to Poynting, undated but between 12 and 18 Oct. 1888; Lodge added, "I certainly do not know now where [Thomson] is." Heaviside called Thomson's paper "absurdly behind the times"; see Heaviside notebook 3a, entry 164, OH-IEE.

42. William Thomson 1888.

have been to some extent not altogether wrong," produced "some little amusement" at the meeting, and the *Times* singled out Thomson's apparent change of heart as an event "of great importance to all interested in electro-magnetic theory."[43] The news of Hertz's discovery had clearly had its effect on Thomson, and the Maxwellians were not slow to exploit the opening it provided.

The discussion of Thomson's paper started immediately and ran throughout the meeting. "Long-standing students of both Thomson and Maxwell" joined in, Lodge reported, and engaged in "keen competition to see through the fog of symbols, to disentangle their real physical meaning, and to find a right and proper position for the half lopped-off apparent excrescences of Maxwell." The debate soon came to focus on ψ and whether it should be dropped from the propagation equations. "Many deaths, and many bringings to life, this unfortunate symbol suffered in the course of the week," Lodge said, and a consensus was slow to emerge.[44] All acknowledged the treatment of ψ and the associated quantity J in Maxwell's *Treatise* to be unsatisfactory, and the Maxwellians found themselves criticizing Maxwell's book in order to defend his theory, a tactic with which they were already becoming familiar. Thomson could be seen going around the meeting "with the second volume [of Maxwell's *Treatise*] under his arm, every now and then appealing to FitzGerald to explain a passage," and Rayleigh, Lodge, and Rowland were also active in the floating seminar that filled the week at Bath.[45]

The propagation of ψ was one of the points FitzGerald had puzzled over with his wheel-and-band model (see Chapter 5), and he had solved it then by assigning a new value to J, the divergence of the vector potential $\mathbf{A}$. Heaviside had tried the same route before deciding to abandon the potentials in favor of the electric and magnetic forces $\mathbf{E}$ and $\mathbf{H}$, but his work had as yet attracted little attention. It was over the weekend at Bath that FitzGerald went back through his notebook and scrawled beside his new equation linking J and ψ the words: "very important. 9.9.88. Must be all in O. Heaviside."[46] At the meeting he went out of his way to say, "Most of these refined points would probably be found mathematically treated somewhere in the writings of Mr. Heaviside," and Heaviside's papers, previously almost ignored, soon began to attract much wider notice.[47]

The propagation debate reached its climax at Bath on 12 September,

43. *Nature 38* (1888): 469; *Times*, 14 Sept. 1888, p. 6.
44. Lodge 1888e: 624.
45. Lodge 1931a: 98–99; Lodge 1888e: 624.
46. FitzGerald notebook 10376, opp. p. 37, FG-TCD Library.
47. Lodge 1888e: 625.

the last day of the meeting, when Henry Rowland was scheduled to give his paper "On a Modification of Maxwell's Equations of Electromagnetic Waves."[48] FitzGerald, who was chairing the session, had announced that Rowland was going to "murder ψ" by showing once and for all that the electrostatic potential had no place in the basic propagation equations. Because of a scheduling mix-up, Rowland was late in arriving, and those present asked FitzGerald to fill the time by reporting on the current status of the propagation debate. He demurred, saying he did not want to intrude on Rowland's paper, and talked instead about the vortex sponge theory, but it was clear that interest in the fate of ψ was running high. When Rowland finally arrived, he "stated his case briefly and with hesitation," according to the report in *Engineering*, "as Professor Fitzgerald knew more about it than he did himself." A lively discussion ensued, *Engineering* said, in which "everybody expressed regret at the absence of Mr. Heaviside, and kept on his guard."[49]

Gradually, however, FitzGerald and Rowland were able to win general agreement that ψ should indeed be murdered, at least in treating propagation. Lodge summarized their reasoning in his "Sketch" of the meeting:

> Plainly, an electrostatic field cannot arise without the motion of some electricity, either with or through a conductor. Now, whenever electricity moves . . . its motion generates a magnetic field. When the motion ceases the field at once subsides, and in subsiding it may produce a succession of diffusing and dying away induction currents in neighbouring conductors; or it may, if the circuit be an incomplete one, leave a permanent vestige of itself in the dielectric as a field of strained ether—this state of strain being what we call electrostatic potential, and the field being familiar to us as an electrostatic field. Generating it in this way, all distinction between rate of propagation of electrostatic and electromagnetic potential vanishes, they both travel together with the velocity of light; or rather, the thing which travels is the magnetic potential, and its permanent effect *in situ* is the electrostatic potential.[50]

ψ was thus banished from the propagation equations, though the electromagnetic or vector potential **A** still retained a prominent place in them.

Soon, however, **A** came under attack as well. After reading Lodge's account of the propagation debate in the *Electrician*, Heaviside declared it had not gone far enough; his own goal, he told Lodge, was "not merely

48. Only the title of Rowland's paper appears in *BA Report* (1888), p. 617.
49. *Engineering 46* (1888): 352.
50. Lodge 1888e: 624–25.

the murder of Maxwell's ψ, but of that wonderful three-legged monster with a scalar parasite on its back, the so-called electrokinetic momentum at a point!"—that is, the vector potential itself.[51] In October he sent the *Philosophical Magazine* a note, "On the Metaphysical Nature of the Propagation of Potentials," in which he sought to clarify the issues raised at Bath and to give an account of his own approach to the problem. Maxwell's reliance on the potentials had been a great mistake, Heaviside said; they did not represent the actual state of the field and so could be used to treat propagation phenomena only in a roundabout way that was "as complex and artificial as it is useless and indefinite."[52] The physical state of the field, in particular the density and flux of energy within it, was determined by the electric and magnetic forces **E** and **H,** and it was these, Heaviside said, that should be the proper objects of attention. "It is perfectly obvious that in any case of propagation . . . it is a physical state that is propagated"; thus it is just as obvious that "it is **E** and **H** that are propagated," not **A** and ψ, he maintained. Once **E** and **H** were installed as the fundamental electromagnetic quantities, as in the new set of "Maxwell's equations" Heaviside had introduced in 1885, the problem of the propagation of the potentials simply evaporated; for it was revealed to be a purely "metaphysical" question, given that all observable phenomena depend solely on the electric and magnetic forces, not on the potentials.[53] This view of the potentials as little more than mathematical fictions gained increasing adherence after 1888 and soon became a central feature of the new Maxwellian synthesis.

Heaviside's attack on the potentials was aimed mainly at Thomson, who had been their chief defender at Bath. Thomson had reportedly said it would be better "to spare ψ, and rather to sacrifice *J*," and after the meeting he continued to argue that the propagation of the potentials was not a purely "metaphysical" question.[54] Indeed, Lodge and others were concerned that he might raise some new objection to the Maxwellians' position and derail their carefully constructed consensus on the nature of electromagnetic propagation.[55]

But this did not occur, and the Bath meeting in fact inaugurated a period in which Thomson was widely regarded as having undergone what *Nature* called "his long-deferred conversion to Maxwell's elec-

51. Heaviside to Lodge, 9 Oct. 1888, OJL-UCL.

52. Heaviside 1892, 2: 484 [1888].

53. Ibid., p. 483.

54. *Engineering 46* (1888): 352; William Thomson to Heaviside, 4 Nov. 1888, OH-IEE, quoted in Heaviside 1892, 2: 490 [1888].

55. Lodge 1888e: 624.

tromagnetic theory of light."[56] After softening his criticism of the theory
at the meeting itself, he expressed an even more favorable opinion of it
in the months that followed. In his presidential address to the Institution
of Electrical Engineers in January 1889, Thomson declared that Maxwell's theory marked "a stage of enormous importance in electro-magnetic doctrine," and in his 1893 preface to Hertz's *Electric Waves,* he
praised Maxwell's "splendidly developed theory" and the way it had
been "established on the sure basis of experiment" by Hertz.[57] Moreover, in 1889 and 1890, Thomson published papers on the structure of
the ether in which he abandoned the ordinary elasticity assumed in earlier purely optical theories and instead ascribed to it the rotational elasticity required for an electromagnetic medium—thus demonstrating,
Heaviside said, that he had indeed "got a proper grip of Maxwell."[58] All
in all, Thomson's reaction was a far cry from his curt dismissal of Maxwell's theory in 1884.

Thomson's move toward Maxwellianism in the fall of 1888 was clearly
a significant transition in his thinking, but it has been minimized or
missed altogether by previous writers—perhaps understandably, since it
was so short-lived.[59] Thomson returned to his old views on the primacy
of the elastic solid within a few years, and letters he wrote to FitzGerald
in 1896 show that by then he no longer accepted even the basic tenets of
Maxwell's theory, particularly concerning electromagnetic propagation.
By the time his *Baltimore Lectures* were finally published in 1904, Thomson was attacking Maxwell's theory as vigorously as when he had first
delivered them some twenty years before.[60]

The Maxwellians watched the evolution of Thomson's views with interest and then with growing dismay. As Heaviside observed to Fitz-
Gerald in 1896,

> Lord K. has the defects of his qualities. He is powerful minded, and has
> devoted so much attention to the elastic solid, that it has crystallized his
> brain. Some years ago he forced his brain into another arrangement. That
> he actually did it (all the harder because of his power) he proved by his
> papers. If he had kept on at it, it might have been a permanent change, but

56. "Electrical Notes," *Nature 39* (1889): 380; see also Heaviside's remark to Lodge, 21
Sept. 1888, OJL-UCL: "Sir W. has been converted at this last B.A. meeting, I see."
57. William Thomson 1882–1911, 3: 490 [1889]; Kelvin, "Preface," in Hertz 1893: xiii.
58. Heaviside to FitzGerald, 9 June 1896, FG-RDS; see William Thomson 1882–1911,
3: 462–72 [1889–90].
59. See, e.g., Knudsen 1985: 171–76.
60. William Thomson 1904: 9. The letters Kelvin and FitzGerald exchanged in 1896 are
quoted in Thompson 1910, 2: 1064–72, and discussed in Smith and Wise 1989: 492–94.

it seems he didn't; for his forced readjustment was *unstable,* and he has reverted automatically to old notions.[61]

Hertz's experiments and the Maxwellians' arguments had been strong enough to win Thomson over, Heaviside said, but not strong enough to make the "conversion" stick.

Thomson's opposition to Maxwell's theory before 1888 and again after about 1894 has fostered a picture of him as an unrelenting foe of Maxwellianism. But it is important to recognize that he in fact took a relatively favorable view of Maxwell's theory for a few years after 1888; for without his support or at least acquiescence, it would have been far more difficult for the Maxwellians to establish their theory as the new electromagnetic orthodoxy. Indeed, when Thomson praised Heaviside's theory of cable transmission in his 1889 address to the Institution of Electrical Engineers, his remarks were widely regarded as putting the official seal of approval not just on Heaviside's work but on the Maxwellian program as a whole.[62] Whether ψ was murdered or allowed to live was of direct concern only to a few specialists, but the emergence of the Maxwellians as authorities to whom even Thomson was willing to defer, at least for a time, was an event of broader and more concrete importance. Thomson enjoyed enormous prestige both within and beyond the physics community, and his support for the Maxwellians, brief and qualified as it was, helped them gain an audience and win legitimacy at a crucial time in their history.

"Practice versus Theory"

The Maxwellians' rise to authority in the electrical world was not entirely unopposed, particularly when it began to impinge on the established community of electrical engineers, or "practical men," as they styled themselves. In the late 1880s the tension between the two groups flared into what became known as the "Practice versus Theory" debate, and the Bath meeting was the site of some of its sharpest exchanges.[63] These were partly a continuation of the long-simmering battle between Preece and Heaviside over inductive loading, and partly a renewal of the dispute between Preece and Lodge over proper methods of lightning protection. At one level, both of these controversies concerned the effect of self-induction on oscillatory currents, a subject the experiments of

61. Heaviside to FitzGerald, 12 June 1896, FG-RDS.
62. William Thomson 1882–1911, 3: 487–90 [1889].
63. See Jordan 1982a and Hunt 1983.

Lodge and Hertz had also brought into prominence. But the more fundamental issue was that of authority: who was to *decide* that self-induction was or was not important, the "practical men," or what one of Preece's friends called "the Maxwellian clique?"[64] The battle over this question was to prove as important for the success of the Maxwellian program as any of those waged over deep issues of theory.

Preece set the confrontational tone at Bath in his presidential address to Section G (Engineering) on 6 September, when he accused physicists of being unable to answer even the simple question What is electricity? Engineers had a clear and practical conception of electricity as a form of energy that could be bought, sold, and put to work, he said, but physicists could describe it only as "a vague speculative unreality" somehow tied up with the ether. After sketching the "modern view" of the ether and electricity Lodge had given in his articles in *Nature,* Preece declared, "The practical man, with his eye and his mind trained by the stern realities of daily experience, on a scale vast compared with that of the little world of the laboratory, revolts from such wild hypotheses, such unnecessary and inconceivable conceptions, such a travesty of the beautiful simplicity of nature."[65] The "practical man" would not be dictated to by the theorist, Preece said, and would not let "wild hypotheses" about the action of the ether intrude into practical engineering. He singled out Lodge's "rather fanciful speculation" about the role of self-induction in lightning conductors as an example of the dangers of loose theorizing, noting pointedly, "The whole subject is going to be thoroughly discussed at this meeting."[66]

The promised discussion on lightning conductors was held on Tuesday, 11 September, and it provided the chief arena at Bath for the battle between "practice" and "theory." The officers of Section A had arranged a joint session with Section G as a way to air the issues Lodge had raised in his controversial lectures at the Society of Arts in the spring and so give some spice to the annual meeting. Preece, as the president of Section G and the leading defender of the orthodox methods of lightning protection, was chosen to open the discussion, with Lodge to reply; FitzGerald, as the president of Section A, served as chairman.[67] The session aroused intense interest and drew such a large crowd that it nearly emptied several of the other sections.

64. Edwin Clark to Preece, 22 Sept. 1888, Preece Collection, NAEST 53/27, IEE Archives. Clark had helped pioneer British telegraphy in the 1840s and 1850s and gave Preece his start in the business in 1853; see Baker 1976: 40.

65. Preece 1888: 791.

66. Ibid., p. 782.

67. Preece et al. 1888. This is a nearly verbatim transcript of the discussion and is much fuller than the version in *BA Report* (1888), pp. 591–615.

Caricature of W. H. PREECE (standing) and OLIVER LODGE. From *Electrical Plant*, 1888

_____

"But the discussion had deeper issues than the important subject of the form and distribution of lightning-conductors."

—*Times* report on the 1888 British Association
debate on lightning conductors

The debate was conducted in the grand Victorian manner, and the *Times* reported that "even from the merely oratorical standpoint there were some excellent speeches" on both sides.[68] Preece and Lodge tossed back and forth a chaffing reference to "Balaam's ass" and made other humorous remarks but also traded some hard blows. Preece opened by attacking all of Lodge's principal claims: that lightning discharges were oscillatory, that self-induction was thus brought into play, and that it acted to confine currents to the outer skin of the conductor. He called self-induction "a bug-a-boo" and expressed dismay at "the way in which lately self-induction had been brought in to account for every unknown phenomenon." Lodge and his allies were in the grip of a "mania . . . for self-induction," Preece said, and their enthusiasm for "mere mathematical development" was leading them far astray.[69] "Practical men" like himself, by contrast, were guided only by the sure light of experience, and Preece boasted that "he made mathematics his slave, and he did not allow mathematics to make him its slave." Indeed, he heaped so much opprobrium on the heads of mathematicians that Lord Rayleigh finally rose to protest that "during Mr. Preece's address one felt that 'mathematician' was becoming almost a term of abuse."[70] Rayleigh and Thomson joined in defending the value of Lodge's theory and experiments, and according to the *Times*, "It was generally agreed that Professor Lodge had the better of it" in the debate. But it was clear, the *Times* went on, that "the discussion had deeper issues than the important subject of the form and distribution of lightning-conductors."[71]

One did not have to dig very far to find that one of those issues concerned Oliver Heaviside and his theory of the propagation of currents. Repeatedly since the spring of 1888, Lodge had cited Heaviside's work in support of his theory of lightning conductors, and in a 17 August letter to the *Electrician*, Heaviside had reciprocated by drawing attention to the way Lodge's experiments confirmed his claims about the role of self-induction in the transmission of signals.[72] Preece's awareness of the close connection between Lodge's work and Heaviside's may well have contributed to the sharpness of his attacks on Lodge at Bath. Certainly Heaviside thought so. After reading the report of the lightning conductor debate, he told Lodge he was sure "Mr. P.'s very strong attack upon you is, in some parts, directed against me"; in particular, "the diatribes against mathematicians are, unless I am greatly mistaken, meant for

68. *Times*, 14 Sept. 1888, p. 6.
69. Preece et al. 1888: 646.
70. Ibid., pp. 645, 674.
71. *Times*, 14 Sept. 1888, p. 6.
72. Heaviside 1892, 2: 486–88 [1888].

me."[73] Heaviside was especially struck by a passage near the end of the debate in which Preece had said that his criticisms of mathematicians were not aimed at such men as Rayleigh and Thomson. Rather, he said, the problem was that

> in the present day, what with Technical Institutes, what with the innumerable students that were sent into the world, especially into the electrical world, every year, they found some of those young fellows coming out every year with a smattering of mathematics; they wrote Papers for the technical journals, and they thrust upon the electrical world conditions and conclusions arrived at by their mathematics with a coolness and an effrontery that were simply appalling. It was to try and check that sort of thing, that was growing to a very serious extent, that he had made the few remarks he had.[74]

This was a good summary of the fear that actuated Preece's attacks—the fear that electrical engineering was about to be taken over by young men trained in mathematical theory who would push Preece and the rest of the old-style "practical men" aside. Moreover, it seemed to Heaviside that Preece's remarks were aimed directly at him; for though he had not learned his mathematics at a technical institute, he was otherwise just the sort of theorist Preece had said he wanted to "try and check."[75]

Heaviside had looked upon Lodge as an ally from the time he first read his lightning lectures, and the debate at Bath, in which Lodge had faced and bested Heaviside's great enemy, Preece, strongly reinforced this feeling. Immediately after the meeting, he sent Lodge three long letters detailing Preece's sins and seeking Lodge's aid in Heaviside's campaign to strike back.[76] Lodge took Heaviside's charges against Preece very seriously; he had come to regard Heaviside as a fellow Maxwellian and an important source of support for his own contentions, and he was especially alarmed by the accusation that Preece had blocked the publication of some of Heaviside's papers. Lodge was hesitant to act publicly after hearing only Heaviside's side of the story, but he offered to write privately to Rayleigh or to the editor of the *Electrician* on Heaviside's behalf.[77]

Whether at Lodge's prompting or not, early in October the *Electrician* published an editorial entitled "Practice versus Theory." In the course of an attempt to cool the conflict between the theorists and the "practical

73. Heaviside to Lodge, 24 Sept. 1888, OJL-UCL. Preece made a point of heaping scorn on the word *impedance*, which he must have known had been coined by Heaviside.
74. Preece et al. 1888: 679.
75. See Heaviside notebook 3a, entry 157, OH-IEE.
76. Heaviside to Lodge, 17, 21 and 24 Sept. 1888, OJL-UCL.
77. Lodge to Heaviside [draft], 23 Sept. 1888, OJL-UCL.

men," the editors mentioned the dispute between Heaviside and Preece over signal propagation and suggested "it may yet prove that Mr. Oliver Heaviside's far more complex theory approximates nearer to the ultimate truth."[78] This seemingly innocent remark gave Heaviside the opening he had been waiting for, and he proceeded to launch a blistering attack on Preece and the "practical men." In a letter entitled "Practice versus Theory: Electromagnetic Waves," he accused Preece of being wholly ignorant of the nature and effects of self-induction and of the most elementary points of scientific procedure. Preece had said that self-induction caused retardation and that its effects were invariably pernicious; Heaviside replied that he would be glad to show, if the *Electrician* would give him space, that in fact "the despised self-induction is the great moving agent" by which the signals were carried along and that it was thus of positive benefit. He pointed out how recent scientific and technical advances, particularly the experiments of Hertz and Lodge with waves on wires, tended more and more to confirm his views. "Self-induction," he declared,

> in spite of strenuous efforts to stop it, goes on moving; nay, more, it is accumulating momentum rapidly, and will, I imagine, never be stopped again. It is, as Sir W. Thomson is reported to have remarked, with a happy union of epigrammatic force and scientific precision, "in the air." Then there are the electromagnetic waves. Not so long ago they were nowhere; now they are everywhere, . . . Now these waves are also in the air, and it is the "great bug" self-induction that keeps them going.[79]

He was even moved to burst into verse, at least in private:

> Self-induction's "in the air,"
>   Everywhere, everywhere;
> Waves are running to and fro,
>   Here they are, there they go.
> Try to stop 'em if you can
>   You British Engineering man![80]

Heaviside, as it turned out, was right: self-induction and all that went with it did prove unstoppable. Maxwell's theory had been so strengthened by the experiments of Hertz and Lodge, and the ground of practical engineering so well prepared by technological progress, that it proved impossible to exclude advanced theory from practical applica-

---

78. *Elec. 21* (1888): 730–31.
79. Heaviside 1892, 2: 488–90 [1888]; cf. Heaviside notebook 3a, entry 157, OH-IEE. For Thomson's remark that "self-induction was in the air," see *Times,* 13 Sept. 1888, p. 11.
80. Heaviside notebook 7, p. 113, OH-IEE; quoted (with some minor errors) in Sumpner 1932: 838n.

tions for long after 1888. This is not to say that the practice versus theory debate ended abruptly as soon as electromagnetic waves were discovered; some of the sharpest exchanges came late in 1888 and in 1889, and the debate dragged on for two or three years after that.[81] But claims of the kind Preece had made at Bath, that self-induction was a "bug-a-boo" and that mathematical theory was of no practical use to anyone, were rapidly becoming untenable. There was growing acceptance not only of Heaviside's dictum that it was "the duty of the theorist to try to keep the engineer . . . straight" but also of the principle that the engineer had a duty to heed the theorist on matters of basic electrical doctrine.[82]

In an anonymous note in the *Electrician* soon after the Bath meeting, Lodge had remarked, "Usually the only cure for the 'practical' generation is to die off, and for a new generation to arise."[83] This was harsh, but it was essentially what happened: as academically trained electrical engineers began to appear in Britain toward the end of the 1880s, many from programs started by Maxwellians, they pushed the old-style "practical men" aside, particularly in projects involving the new and more complex phenomena encountered in long-distance telephony and alternating current power distribution. By the mid-1890s ultimate authority on electrical questions had passed almost entirely to the scientists—in particular, to the Maxwellians.

The Bath meeting marked a watershed in the fortunes of the Maxwellians and their program. The work FitzGerald, Lodge, Heaviside, and Poynting had done in clarifying and extending Maxwell's theory had attracted relatively little notice before 1888; to those outside the narrow circle of British electromagnetic specialists, their work often appeared to be mere paper theorizing, with little bearing on experimental facts or practical applications. Hertz's experiments changed all this; as Heaviside later said, "The very slow influence of theoretical reasoning on conservative minds was enforced by the common-sense appeal to facts," and the Maxwellians were suddenly able to attract the attention and compel the assent of those who were unswayed by the mathematical beauties of their theory.[84] As they emerged from the Bath meeting, the Maxwellians were in an excellent position to consolidate their gains and strengthen their influence in the scientific world.

81. See Hunt 1983 and Nahin 1988: 162, 173–74.
82. Heaviside 1892, 2: 488 [1888].
83. *Elec. 21* (1888): 749. On Lodge's authorship, see Heaviside notebook 3a, entry 157, OH-IEE. Lodge tried to soften his remarks, without acknowledging them as his own, in a subsequent letter to the editor; see Lodge, "Practice v. Theory," *Elec. 21* (1888): 800.
84. Heaviside 1893–1912, 1: 308 [1892].

CHAPTER EIGHT

The Maxwellian Heyday

The rapid rise in the prestige and acceptance of Maxwell's theory which followed the Bath meeting was paralleled by—indeed, was virtually inseparable from—the Maxwellians' own rise to leadership in the community of electrical specialists. Moreover, as the Maxwellians rose, they drew more closely together, so that within a few months after the announcement of Hertz's discoveries they formed a much stronger and more cohesive group than they ever had before. FitzGerald and Lodge, who had long formed the core of the group, were joined in 1888 by both Heaviside and Hertz, and in the succeeding years these men, along with several lesser figures, cooperated closely in the explication of what Heaviside called "that mine of wealth 'Maxwell.' "[1] It was between 1888 and 1894 that the Maxwellians were most united as a group, that their ideas were brought to their fullest and purest development, and that their influence in the scientific world was at its height. It is in this period that we can best see just how the Maxwellian group worked and how the differing personalities and talents of its members meshed in a single cooperative enterprise. By the mid-1890s they had succeeded in promulgating a new and distinctively Maxwellian theoretical synthesis, had established it as a new scientific orthodoxy, and had begun to extend it into regions that touched the boundaries of classical physics. The triumph of the Maxwellian program seemed almost complete.

1. Heaviside 1892, 2: 393 [1888].

Strengthening the Links

FitzGerald and Lodge had already been friends for nearly ten years when the discovery of electromagnetic waves suddenly put them and their work at the center of scientific attention. They became even closer, both personally and professionally, between 1888 and 1894, exchanging visits to Dublin and Liverpool and seeing each other often at meetings of scientific societies in London and of various university examining boards. Their increasing intimacy was reflected in their correspondence: after the summer of 1890 they jokingly signed their letters to each other with Greek initials—"φ" for FitzGerald, "Λ" for Lodge.[2] Over two hundred of these letters survive, and they show by their content that the two men were close collaborators, and by their tone that they were the closest of friends.

Both FitzGerald and Lodge carried on their careers with little outward change in these years, continuing their teaching and research and establishing themselves more firmly both locally and in the broader scientific community. Each soon gathered a group of junior colleagues— John Joly, Frederick Trouton, and Thomas Preston around FitzGerald in Dublin; A. P. Chattock and J. L. Howard with Lodge in Liverpool— who made important contributions to their work, especially on the experimental side, and helped spread Maxwellian doctrines more widely.[3] Even more important, however, were the ties they established almost simultaneously in 1888 and early 1889 with both Oliver Heaviside and Heinrich Hertz. With these ties, the final links in the Maxwellian chain were joined and the Maxwellian heyday began.

Hertz's entry into the Maxwellian circle was especially dramatic, as we have seen, and he soon became one of its most prominent members. When FitzGerald wrote him in June 1888 after first hearing of his propagation experiments, the two men began a correspondence that continued for several years and became quite intense early in 1889. Lodge also began to correspond with Hertz in 1888 (though most of his letters were very brief) and, like FitzGerald, to exchange offprints with him. Many of the letters from both FitzGerald and Lodge concerned how Hertz's experiments could be repeated and extended and their significance for Maxwell's theory made better known, and all three men worked together to publicize the new discoveries. When Lodge lectured on the discharge of a Leyden jar at the Royal Institution in March 1889,

---

2. Lodge 1931b: 143.
3. On Joly and Trouton, see their *DSB* entries; on Preston, see Weaire and O'Connor 1987; on Chattock and Howard, see Lodge 1931b: 148–49, and Rowlands 1990: 18–20 and 58–59.

OLIVER LODGE AND G. F. FITZGERALD (with unidentified woman and young man), 1890s. Courtesy University of Birmingham Library

―――――――

"It is right also that he should be lamented on his human side by one who loved him as a brother."
     —Oliver Lodge, in obituary of G. F. FitzGerald, March 1901

he made a point of drawing attention to the related work of both Hertz and FitzGerald. He even suggested to his audience that "a discourse on Hertz's researches from Prof. Fitzgerald next year would be not only acceptable to you, but highly conducive to the progress of science," and in March 1890, FitzGerald duly spoke at the institution. In his talk, "Electromagnetic Radiation," he repeated many of Hertz's experiments and showed several new ones of his own.[4]

Lodge also took the lead in inviting Hertz to the 1889 and 1890 meetings of the British Association, but despite the promise of "a most cordial welcome," Hertz had to decline, in 1889 because he was busy preparing an important address to the German Congress of Natural Scientists (a sign of the growing recognition of his work in his own country), and in 1890 because of his required military service.[5] But when the Royal Society awarded Hertz its prestigious Rumford Medal in November 1890, he was able to go to London to accept it in person and there met with Lodge, FitzGerald, Poynting, and many other prominent British physicists.[6] This was an important event for the Maxwellians; for in recognizing Hertz's work, the Royal Society had implicitly put its stamp of approval on the Maxwellian program as a whole. Hertz was very well received in London (he was, he later said, almost worn out by the round of receptions and dinners held in his honor), and his visit helped cement the bonds between himself and the British Maxwellians.

Unfortunately, Hertz was not able to remain an active member of the group for very long. His move early in 1889 from the Technische Hochschule at Karlsruhe to the chair of physics at the University of Bonn was a definite step up the academic ladder, but it was a step down in the quality of laboratory facilities and left him with relatively few opportunities for experimental work.[7] As before at Kiel, Hertz shifted his attention to theoretical studies, producing two important papers on the fundamental equations of electrodynamics and beginning work on a

4. Lodge 1889a, repr. Lodge 1889b: 369; FitzGerald 1902: 266–76 [1890].

5. Lodge to Hertz, 24 June 1889 and 29 Apr. 1890, HH-DM; Hertz to Lodge, 21 July 1889 and 1 May 1890, OJL-UCL. See also O'Hara and Pricha 1987: 92–97 and Hertz 1977: 293–97.

6. Hertz 1977: 307–11; O'Hara and Pricha 1987: 109–19. Hertz was in London from 29 Nov. to 2 Dec. and made a brief trip to Cambridge on 3 Dec. to meet J. J. Thomson and see the Cavendish Laboratory.

7. Jungnickel and McCormmach 1986, 2: 93–95. Friedman 1989: 13–15 quotes Vilhelm Bjerknes, Hertz's research student in 1890, on the inadequacies of the Bonn laboratory; Hertz also suffered from eye problems incurred during his earlier spark experiments and could not make further such observations for some time. See also Hertz 1977: 283, 289, 301.

new formulation of the principles of mechanics, in which he hoped to lay the groundwork for a comprehensive physics of the ether. Soon, however, he fell ill with a serious infection that spread to his jaw and sinuses. "At present my nose is my universe," he told his parents in August 1892, "and all my work is remote and meaningless to me."[8] His health improved enough in 1893 for him to complete his new *Principles of Mechanics*, but the illness soon returned.[9] After a long and painful struggle, Hertz died on 1 January 1894 at the age of thirty-six.

Hertz made an enormous contribution to the Maxwellian program, but his passage through the Maxwellian group was all too brief. From a longer perspective, the addition to that circle of Oliver Heaviside, also in 1888, was perhaps more important. Heaviside had been exploring and revising Maxwell's theory on his own since the early 1880s, but it was only after he linked up with the other Maxwellians late in the decade that his work began to attract much notice. As an isolated individual without rank or position, Heaviside could easily be dismissed as a mere crank; as a member of the active and prominent Maxwellian group, his ideas commanded more serious attention and soon gained wide acceptance.

Like Hertz, Heaviside first came into sustained contact with the other Maxwellians in June 1888, when he wrote to ask Lodge about his lightning lectures. Recognizing Heaviside's talent and his potential value to the Maxwellian cause, Lodge quickly became one of Heaviside's first important allies in the scientific world and set out to win public recognition for his work. Lodge was the first to nominate Heaviside for fellowship in the Royal Society; it was to him that Heaviside turned in 1889 for advice on distributing copies of his little book *Electromagnetic Waves;* and it was he who took the lead in persuading Macmillan's to publish Heaviside's collected *Electrical Papers* in 1892.[10] Outwardly, Lodge and Heaviside had little in common; the busy and very public professor in Liverpool led a life very different from that of the unemployed ex-telegrapher in Camden Town, a difference that grew even wider after Heaviside moved with his parents to the little Devonshire town of Paignton in September 1889. But their talents and personalities were complementary rather than conflicting, and though they were to meet only once, Lodge and Heaviside soon became friends as well as allies.

Lodge was not enough of a mathematician to discuss much of Heaviside's theoretical work with him on equal terms, and only a few of their

8. Hertz 1977: 327.
9. Hertz 1899 [1894].
10. Heaviside to Lodge, 19 and 30 Jan. 1889 and 15 and 23 Mar. 1891, OJL-UCL.

nearly 150 surviving letters deal with the higher reaches of Maxwellian theory. But Lodge had great respect for Heaviside's abilities and came to regard him as the leading authority, along with FitzGerald, on the mathematical interpretation of Maxwell's theory. Heaviside in turn held Lodge's talents as an experimenter and expositor in high regard. When Lodge's *Modern Views* appeared in the summer of 1889, Heaviside wrote to praise it, declaring, "The time is very opportune." "Popularisation of Maxwellian views is much wanted," he said, "and yours is the first of many." He also thanked Lodge for having mentioned him favorably: "I am very glad you have put me in," he said, "because of the relative permanency of books. I am immortalised until the next geological catastrophe comes."[11] Lodge's prominence as a public speaker and writer enabled him to win Heaviside a far wider audience than he would otherwise have attracted, and by 1889, Heaviside was fully aware of the value of such recognition.

The little book *Electromagnetic Waves* that Heaviside arranged to have printed and distributed early in 1889 was part of his deliberate effort to gain wider notice for his work, and the copy he sent to Hertz paid especially rich dividends. Hertz was glad to have the book but sorry that Heaviside had not included his return address. "I know not even, where he lives," Hertz lamented to Lodge, and he asked both Lodge and FitzGerald for Heaviside's address so that he could thank him properly.[12] FitzGerald obliged, and the letter Hertz sent to Heaviside on 12 February 1889 helped complete another important link in the Maxwellian chain. Hertz also sent copies of six of his recent papers, a gift Heaviside especially appreciated since he did not have access to German journals and had previously known of Hertz's work only at second hand. Heaviside responded with a long and important letter in which he explained his vector notation and his approach to Maxwell's theory, particularly his reasons for replacing Maxwell's equations in the scalar and vector potentials with his own symmetrical equations in the electric and magnetic forces.[13]

Hertz thanked Heaviside for these explanations, telling him, "I more clearly understood your methods from your letter than from your book, where they lie hidden beneath a great number of special cases." He strongly endorsed the abandonment of the potentials and told Heaviside he believed "that you have gone further on than Maxwell, and that if he had lived he would have acknowledged the superiority of your

11. Ibid., 20 July 1889.
12. Hertz to Lodge, 7 Feb. 1889, OJL-UCL; FitzGerald to Hertz, 8 Feb. 1889, HH-DM.
13. Heaviside to Hertz, 14 Feb. 1889, HH-DM.

methods."[14] In a later letter, after Heaviside had further explained his terminology and methods, Hertz wrote:

> The fact is that the more things became clearer to myself and the more I then returned to your book, the more I saw that essentially you had already made much earlier the progress I thought to make, and the more the respect for your work was growing in me. But I could not take it immediately from your book, and others told me that they could hardly understand your writing at all, so I felt obliged to give you warning that you are a little obscure for ordinary men.[15]

Hertz and Heaviside's correspondence continued to be quite lively through the rest of 1889 and into 1890, often touching on such difficult areas of Maxwellian theory as the true nature of force and displacement and the proper form of the field equations in a moving medium. The two men never met, though Hertz apparently gave some thought to making a side trip to Paignton during his 1890 visit to London. Heaviside was "shocked" that Hertz would even think of going two hundred miles out of his way just to meet him; "There are far more important persons to see without going so far," he said, though he added that if Hertz ever happened to be in South Devon, "I should be delighted to have the chance of making your acquaintance in the flesh."[16]

Hertz's theoretical writings, particularly his 1890 paper "On the Fundamental Equations of Electrodynamics for Bodies at Rest," played a major role in establishing the four symmetrical equations connecting the electric and magnetic forces as the standard form of "Maxwell's equations." Arnold Sommerfeld, then a young physics student at Königsberg, later described how deeply he and his contemporaries were impressed by the clarity and coherence of Hertz's "purified" version of Maxwell's theory. When he read the 1890 paper, "It was as though scales fell from my eyes," Sommerfeld said, and he followed Hertz's lead closely in his own later work on electromagnetism.[17] Hertz was usually credited on the Continent with having originated this form of the equations and with having led the shift away from regarding the potentials as fundamental quantities—and he had indeed published the four equations (in Cartesian form) as early as 1884, though with a very shaky physical justification.[18] But in light of Hertz's close study of Heaviside's papers and

14. Hertz to Heaviside, 21 Mar. 1889, quoted in Appleyard 1930: 238. Hertz's letters to Heaviside have apparently been lost except for the excerpts quoted by Appleyard and reprinted in O'Hara and Pricha 1987.
15. Hertz to Heaviside, 5 May 1889, in Appleyard 1930: 238.
16. Heaviside to Hertz, 8 Dec. 1890, HH-DM.
17. Sommerfeld 1952: 1–2.
18. Hertz 1896: 273–90 [1884]; D'Agostino 1975: 286–92.

especially his letters on precisely this point early in 1889, it seems clear that, in his final reformulation of "Maxwell's equations," Hertz was strongly influenced by Heaviside's work.

Heaviside certainly thought so, pointing to "the great contrast between his early and his later methods, without natural conjunction or intermediate stages" as evidence that Hertz "had taken in more than he himself suspected" from Heaviside's writings and had jumped directly from his confused earlier ideas to the final form Heaviside had given.[19] Hertz did not mention his own 1884 paper when he returned to the equations in 1890, and he explicitly credited Heaviside with priority in having recast Maxwell's equations and explained their proper meaning and use.[20] After Hertz's death, many physicists spoke of "the Maxwell-Hertz equations"; Einstein, for example, did so in his famous relativity paper of 1905.[21] But without wishing to take anything away from the memory of Hertz, whom he regarded as a friend, Heaviside was quite insistent that the equations should properly be ascribed to "Maxwell, Heaviside, and Hertz," and an examination of Hertz's own writings, particularly his correspondence with Heaviside, suggests that this name would indeed better reflect the actual course of influence in the reshaping of "Maxwell's equations."[22]

There was one more major link in the Maxwellian chain, that between FitzGerald and Heaviside, and it was in many ways the most important of all. Both were recognized as leading authorities on Maxwell's theory, and in their extensive correspondence, running from late 1888 to FitzGerald's death in 1901, they dealt with each other as equals, often delving into the most obscure and recondite regions of Maxwellian theory. They discussed the nature of charges and currents, the propagation of waves across space and along wires, and the relationship between the structure of the ether and the mathematical formalism of Maxwell's theory; and they returned again and again to one of the most difficult and fundamental questions in Maxwellian theory, that of the nature of the electromagnetic field around a moving charge. It was largely through these exchanges that the full Maxwellian synthesis took shape in the early 1890s, and through their efforts that it was disseminated in Britain and eventually throughout the world.

FitzGerald had great respect for Heaviside's work and played a key role in bringing it to the attention of other scientists. The review of

    19. Heaviside to Lodge, 6 May 1889, OJL-UCL.
    20. Hertz 1893: 196–97 [1890].
    21. Einstein 1905, in Miller 1981: 404.
    22. See Heaviside to FitzGerald, 12 July 1897, FG-RDS, and August Föppl to Heaviside, 30 June and 22 July 1897, OH-IEE.

Heaviside's *Electrical Papers* he wrote for the *Electrician* in August 1893 was especially strong and influential and was later cited as evidence of how high Heaviside's work stood in the eyes of those competent to judge.[23] Heaviside himself told FitzGerald that the review was "too too" and that he would "have to be taken down a few pegs," but he could not hide his pleasure at the praise FitzGerald had heaped on him.[24] "Since Oliver Heaviside has written," FitzGerald declared, "the whole subject of electromagnetism has been remodelled by his work"; with his "extraordinarily acute and brilliant mind," Heaviside had cleared the way for the creation of a "New Electromagnetism" in which the leading principles of Maxwell's theory would be laid out with unprecedented clarity and comprehensiveness.[25] FitzGerald repeated and extended this praise a year later in a review of the first volume of Heaviside's *Electromagnetic Theory*, parrying the usual objections to Heaviside's difficult mathematical style by saying that while his "analytical engines" might seem unwieldy to the uninitiated, "It is just these extraordinary methods of attack that betoken genius and enlarge our views of Nature."[26] In these reviews, and in innumerable private contacts with other physicists, FitzGerald threw the weight of his own considerable reputation behind Heaviside and so helped spread an appreciation of his writings to an audience that might otherwise have ignored them.

FitzGerald and Heaviside became quite close in the course of the 1890s, though their friendship was conducted almost entirely by mail. "I only saw him twice knowingly, once for two hours, and then again for six hours, after a long interval," Heaviside said shortly after FitzGerald's death, "yet we had a good deal of correspondence at one time, and I seemed to have quite an affection for him."[27] They had begun corresponding in December 1888 when Heaviside wrote to ask FitzGerald about reports of his earlier work on moving charges and convection currents. After they had exchanged several letters on the subject over the next few weeks, FitzGerald, who happened to be passing through London, called on Heaviside at his home on 8 February 1889.[28] (The second of the two visits Heaviside later recalled came in September

23. See, e.g., "Oliver Heaviside," *Elec.* 37 (1896): 346–47. This was apparently written by FitzGerald, though the paragraphs praising him and quoting from his review of Heaviside were presumably added by W. G. Bond, the editor of the *Electrician*; see Bond to FitzGerald, 8 June 1896, FG-RDS, asking FitzGerald to write such an article.

24. Heaviside to FitzGerald, 12 Aug. 1893, FG-RDS.

25. FitzGerald 1902: 292–300 [1893].

26. [FitzGerald] 1894.

27. Heaviside to John Perry, [Feb. 1901], quoted in Larmor 1901a, in FitzGerald 1902: xxvii; cf. Larmor to Heaviside, 6 Mar. 1901, OH-IEE.

28. FitzGerald to Heaviside, 27 Dec. 1888, OH-IEE; on FitzGerald's visit, see FitzGerald to Hertz, 8 Feb. 1889, and Heaviside to Hertz, 14 Feb. 1889, HH-DM.

1898, after he had moved to South Devon.) Lodge, too, first met Heaviside early in 1889, visiting him in what he later described as Heaviside's "dismal lodgings at Kentish Town" on 10 March.[29] Lodge had lived in the area in the 1870s and presumably knew his way around, but Heaviside gave detailed directions to his house on St. Augustine's Road, noting that their friend from "the Sister Isle" (i.e., FitzGerald) had had trouble finding it a few weeks before. It would be best "to come before 4, have a cup of tea, and get on a sociable footing at once," Heaviside said. "It is no use coming in the morning, because I am busy grunting out Dr. Watts' lines, 'You have woke me too soon.' "[30]

These visits from FitzGerald and Lodge were important to Heaviside; for they marked his acceptance as a respectable scientific man and a full member of the Maxwellian circle. It was clearly with some pride that he later told Hertz of his "conversation with an eminent scientific man [FitzGerald] last February."[31] Not long before, he had been ignored, even scorned, by other scientists; now he was entertaining them in his home, however "dismal" it might be.

Neither Heaviside nor his parents enjoyed good health in their London lodgings, and in September 1889 they moved to the seaside town of Paignton, where Oliver's brother Charles had offered them rooms above his music shop. Heaviside had no income of his own at the time and only began earning a few pounds again in January 1891, when the *Electrician* began to carry his long series of articles, titled "Electromagnetic Theory," which ran until 1902. He and his parents lived in circumstances that were, he told Hertz, "sufficiently poor," though with "little of the humility which some think should accompany poverty."[32] Heaviside's scientific friends knew of his financial straits, and early in 1894, FitzGerald, Lodge, and John Perry sought to secure a grant for him from the Royal Society's Scientific Relief Fund.[33] Heaviside was touchy about accepting such "charity," however, and despite FitzGerald's patient pleading in a long series of letters, he eventually rejected the offer. "I am eccentric if

29. Lodge 1925a.
30. Heaviside to Lodge, 5 Mar. 1889, OJL-UCL. Heaviside was quoting from "The Sluggard," one of the *Moral Songs* for children by Dr. Isaac Watts (1674–1748): " 'Tis the voice of the Sluggard; / I heard him complain, / 'You have wak'd me too soon, / I must slumber again.' " See Appleyard 1930: 215 on Heaviside's habit of sleeping late.
31. Heaviside to Hertz, 13 July 1889, HH-DM.
32. Ibid., 13 Sept.
33. FitzGerald initially asked Heaviside whether he would accept an "honorarium" in a letter of 6 Feb. 1894, OH-IEE, which was also signed by Lodge and Perry. This letter had been carefully crafted (in consultation with Arthur Heaviside) to avoid offending Heaviside's pride; see two earlier drafts among Perry's letters to FitzGerald, FG-RDS. See also Perry to FitzGerald, 13 Feb. 1894, FG-RDS; Perry had found that Heaviside's income for 1893 was only £51, and that it had been even less in previous years.

you like," he told FitzGerald, "and my independence varies inversely as my income"; to preserve that independence, he would rather "live upon my means than . . . be dependent upon anything of the nature of compassionate assistance."[34]

FitzGerald thought Heaviside deserved greater recognition and far more abundant rewards than he ever received, and Heaviside in turn held his Irish friend in the very highest esteem. "'We needs must love the highest when we' know him," he wrote in dedicating the third volume of his *Electromagnetic Theory* to FitzGerald's memory in 1912, and on hearing of FitzGerald's death in 1901, he declared, "The premature loss of a man of such striking original genius and such wide sympathies will be considered by those who knew him and his work to be a national misfortune."[35] When, in 1896, Perry happened to mention to Heaviside that his own high regard for Heaviside's work derived mainly from the praise of it he had heard from FitzGerald, he was, he told FitzGerald, "simply astonished at [Heaviside's] reply: he said he felt so proud and so honoured at your thinking well of him and he used much affecting language!"[36] Heaviside was often contemptuous of those he thought less gifted than himself, but in FitzGerald he recognized a man who was at least his equal, and he was pleased and proud to have won his approval and friendship.

By the spring of 1889, the Maxwellian group had taken solid shape, and its members had established the pattern of interaction they were to follow over the next several years. They exchanged ideas, they became friends, and by publicizing and praising each other's work, they helped advance each other's careers. These levels of interaction—intellectual, social, and professional—were not kept separate, and their fruitful combination in the years after 1888 became one of the keys to the Maxwellians' success.

The Origins of the FitzGerald Contraction

The inner workings of the Maxwellian group can best be illustrated by exploring a particular episode in some detail. An especially clear and interesting case is that of the origin of the "FitzGerald contraction" hypothesis, which states that moving bodies shrink slightly along their line

---

34. Heaviside to FitzGerald, 10 and 13 Feb. 1894, filed among Perry's letters to FitzGerald, FG-RDS. Except for these two letters, most of the exchange between FitzGerald and Heaviside is published in Nahin 1988: 170–73.

35. Heaviside 1893–1912, 3: 89n and "Dedication."

36. Perry to FitzGerald, 7 Mar. 1896, FG-RDS.

of motion through the ether. Though this idea is now generally thought of as part of Einstein's theory of relativity, it was in fact first formulated by FitzGerald in the spring of 1889. By examining not just the published papers but also the letters and even the conversations of the principal figures in this episode, we see that the contraction hypothesis was not just a bright idea hit upon in a vacuum, as it has often been depicted, but was in fact the product of lively exchanges among FitzGerald, Heaviside, and Lodge, with several others playing important indirect roles. The interaction and cooperation that marked the origin of the FitzGerald contraction typified the Maxwellian group throughout the late 1880s and the 1890s; their work was the product not of any single mind but of the interplay of several.

The FitzGerald contraction episode, like so many others in the Maxwellian story, had its start in the "murder of ψ" debate at Bath, in which FitzGerald's contention that all electromagnetic effects are propagated at the speed of light had run up against William Thomson's defense of the older view that the electric potential ψ might well be propagated instantaneously. Lodge's account of the discussion in the *Electrician* prompted Heaviside to send a note "On the Metaphysical Nature of the Propagation of the Potentials" to the *Philosophical Magazine* in October. In it he dismissed the potentials as mere mathematical fictions and declared the question of their rate of propagation to be "metaphysical," since it could not, even in principle, be decided by experiment.[37] He asked the editor of the *Philosophical Magazine* to forward the note to Thomson, one of the journal's advisory editors and the man Heaviside most wanted to impress with his views on the primacy of the electric and magnetic forces.[38] Thomson replied that he did not agree that the rate of propagation of electric potential was "a merely metaphysical question" and suggested that it could be determined experimentally, at least in principle, by moving a charged body back and forth and testing whether sensitive electrometers at different distances from it were affected simultaneously or only after a time lag.[39]

Thomson's letter had important consequences; for it turned the debate over the propagation of electric potential into one about the state of the electromagnetic field around a moving charge. This problem had never before been fully solved; earlier theorists had treated only the slow motion of charges, for which it could be safely assumed that the

37. Heaviside 1892, 2: 483–85 [1889].
38. Heaviside notebook 3a, entries 159 and 164, OH-IEE.
39. William Thomson to Heaviside, 2 Nov. 1888, OH-IEE, part pub. in Heaviside 1892, 2: 490 [1888].

electrostatic field was simply carried along unchanged—that is, that the potential adjusted itself instantaneously. Thomson had in effect challenged Heaviside to find what, on Maxwellian principles, the electromagnetic field would be around a charge moving at *any* speed. Heaviside's answer to this challenge was to play a key role in the genesis of FitzGerald's contraction hypothesis.

Heaviside's first reaction to Thomson's letter was more mundane: he saw it as possible ammunition in his campaign for recognition. Early in November 1888 he asked Thomson for permission to publish his letter on the propagation of the potentials in the *Electrician,* along with Heaviside's reply. He wanted to use Thomson's name as bait; as he said in a postscript to his request, "I may mention as a private reason for wishing to publish it that it may lead the Ed. to see his way to resume my Articles on Propagation, so you might be doing me a great service."[40] Thomson's reputation was enormous; anything from his pen would be snapped up by the editors of the *Electrician.* Heaviside hoped they might be led to swallow his own work as well. Thomson gave his permission, but Heaviside's strategy was only partially successful: the *Electrician* published his letter and shortly thereafter two others on the same subject, but more than two years were to pass before it began carrying his regular series again.[41]

Thomson's challenge led Heaviside back to a question he had first attacked as long ago as 1880 and 1881; indeed, one of his notebooks shows him to have been perhaps the first person ever to treat the moving-charge problem comprehensively on the basis of Maxwell's theory.[42] He had then derived such important results as the force on a charge moving in a magnetic field, $\mathbf{F} = q\mathbf{v} \times \mathbf{B},$ later well known as the "Lorentz force." At that time, however, Heaviside was still using Maxwell's scalar and vector potentials, and his derivations were laborious and cumbersome. He did not then publish his results, many of which were independently derived and first published by J. J. Thomson in April 1881.

J. J. Thomson's paper on moving charges contained several important results and was later often cited as a precursor of electron theory, but it was deeply flawed. Thomson fell into contradictions when he tried to trace all the electromagnetic effects of a moving charge to the displacement currents set up in the surrounding space, and mistakes in his use of the vector potential led him to erroneous values for the energy of and

40. Heaviside to William Thomson [draft], 5 Nov. 1888, OH-IEE.
41. William Thomson to Heaviside, 6 Nov. 1888, OH-IEE; Heaviside 1892, 2: 490–96 [1888].
42. Heaviside notebook 7, OH-IEE, marked "Done in 1880–1."

forces on a moving charge—he gave the Lorentz force, for instance, as half its true value.[43] FitzGerald followed this work closely and corrected the first of Thomson's errors in a note presented to the Royal Dublin Society in November 1881. He pointed out that for all currents to form the closed circuits required by Maxwell's theory, moving charges must themselves be regarded as what he called "convection currents."[44] This constituted a fundamental revision of Maxwell's original equations, albeit a necessary one to keep the theory self-consistent, and FitzGerald's paper, more than Thomson's, contained the real seeds of electron theory. Heaviside finally published his own correction of Thomson's value for the energy of a moving charge in 1885, but it was buried at the end of one of his *Electrician* articles and attracted no attention; Thomson continued to repeat his uncorrected value as late as 1888.[45]

To this point, J. J. Thomson, FitzGerald, and Heaviside had treated only the *slow* motion of charges. They had ignored self-induction, by which the motion of the electric field around a charge induces a magnetic field, whose motion in turn induces an additional electric field, and so on. They had taken into account only the first term of the resulting infinite series—a legitimate procedure as long as the speed of the charge was a small fraction of that of light, but one that tacitly assumed that the field adjusted instantaneously to any motion. William Thomson's challenge to Heaviside on precisely this point in November 1888 drove him to attack the problem again without this simplifying assumption.

Heaviside knew by general reasoning that the electric and magnetic fields around a moving charge would tend to bunch up around the middle, but at first he could derive only a complicated infinite series to describe this effect. To his own surprise, however, he found early in December that this series could be reduced to a fairly simple formula that was exact for any speed up to that of light. At last, he thought, he had the means "to cure Sir W." of his mistaken views on propagation; for this formula clearly showed the transition from the static polar field of an isolated charge to the transverse fields of an electromagnetic wave traveling at the speed of light, with no need for the instantaneous propagation of any potentials.[46] Heaviside recognized that he had found a very fundamental formula, and he immediately sent a letter about it to the *Electrician*, where it first appeared in print on 7 December 1888.[47]

---

43. J. J. Thomson 1881: 241; cf. Whittaker 1951: 306–10.
44. FitzGerald 1902: 102–7 [1881]; also pub. in *Phil. Mag. 13* (1882): 302–5.
45. Heaviside 1892, 1: 446 [1885] and 2: 505–7 [1889]; J. J. Thomson 1888: 32–35.
46. Heaviside to FitzGerald, 30 Jan. 1889, FG-RDS.
47. Heaviside 1892, 2: 494–96 [1888].

Heaviside found that

$$\epsilon E = \frac{q}{r^2} \frac{1 - (v^2/c^2)}{[1 - (v^2/c^2)\sin^2\theta]^{3/2}}$$

directed radially outward, and $H = \epsilon E v \sin\theta$ in circles centered around the line of motion, where ϵ is the permittivity of the medium, q the charge, v its velocity through the ether, c the speed of light, r the distance from the charge to a point, and θ the angle to the line of motion. Note especially the $1 - v^2/c^2$ factor and the way the field lines bunch up around the "equator" (if we picture the charge moving along its polar axis) as the speed increases (see Figure 8.1). This compressed field is in fact just the FitzGerald-Lorentz contraction of the electrostatic field of a charge at rest, in exact accordance with Einstein's theory of relativity. The surface of electrical equilibrium, called a "Heaviside ellipsoid," is an oblate spheroid contracted along the line of motion by a factor of $\sqrt{1 - v^2/c^2}$, although it was not until 1892 that this last point was fully clarified by Heaviside's friend G. F. C. Searle.[48] All of this follows directly from Maxwell's equations and shows quite clearly that "relativistic" effects were already implicit in Maxwell's theory.

Heaviside was proud of his formula, declaring that it was simple and clear enough "to take a place in text-books." He made it the basis of an important paper "On the Electromagnetic Effects Due to the Motion of Electrification through a Dielectric" that he sent to the *Philosophical Magazine* late in December (it was published in April 1889), and he made a point of drawing attention to it in letters to FitzGerald, William Thomson, Hertz, and J. J. Thomson.[49]

It was at just this time and on just these issues that Heaviside began to correspond with FitzGerald—a development of great importance for the origin of the FitzGerald contraction and the evolution of the Maxwellian group. After seeing FitzGerald's early work on convection currents mentioned in the *Electrician* in November, Heaviside wrote to ask him for details. FitzGerald obliged late in December and sent Heaviside offprints of several of his papers.[50] Heaviside took special note of FitzGerald's remark, in his 1881 paper on convection, that it would be "most interesting" to have a solution of the moving-charge problem that was valid for any speed; late in January he wrote to draw FitzGerald's

48. Ibid., p. 514n; cf. Searle to Heaviside, 19 Aug. and 24 Sept. 1892, OH-IEE, and Searle 1896.
49. Heaviside 1892, 2: 495 [1888], 504–18 [1889]. Heaviside to FitzGerald, 30 Jan. 1889, FG-RDS; Heaviside to William Thomson, 27 Feb. 1889, WT-ULC; Heaviside to Hertz, 1 Apr. 1889, HH-DM; J. J. Thomson to Heaviside, 7 Apr. and 26 May, OH-IEE.
50. FitzGerald to Heaviside, 27 Dec. 1888, OH-IEE; cf. Heaviside 1892, 2: 492 [1888].

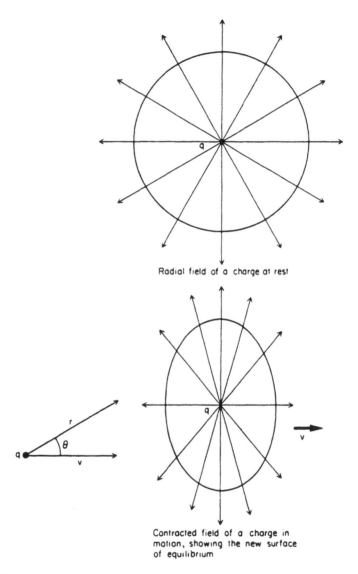

Radial field of a charge at rest

Contracted field of a charge in
motion, showing the new surface
of equilibrium

8.1. The electric field lines around a charge at rest and one in motion, showing the contraction of the surface of equilibrium at high velocity.

attention to the solution he had just published in the *Electrician*. FitzGerald replied that he was "very glad to hear that you have solved completely the problem of the moving sphere."[51] He mentioned its pos-

51. Heaviside to FitzGerald, 30 Jan. 1889, FG-RDS; FitzGerald 1902: 107 [1881]; FitzGerald to Heaviside, 4 Feb. 1889, OH-IEE.

sible relevance to a theory of intermolecular forces and suggested that, as the formula indicated, the velocity of light might be a physical limit to speed.

It was just a few days later, on 8 February 1889, that FitzGerald first visited Heaviside in London. Their main topic of conversation was Hertz's wave experiments, but they apparently discussed the moving-charge problem as well.[52] Their correspondence lapsed for a time after that, but Heaviside made a point of sending FitzGerald a copy of one of his papers on moving charges in the summer of 1889, and the subject came up repeatedly in their letters in the 1890s.[53] In any case, it is clear that by early 1889, FitzGerald knew that Heaviside had shown that Maxwell's theory implies that the electromagnetic field around a moving charge would be compressed by an amount depending on $1 - v^2/c^2$.

The scene now shifts to Liverpool and FitzGerald's good friend Oliver Lodge. Early in 1889 and for several years thereafter, Lodge devoted much of his attention to pondering the motion of matter through the ether and the resulting aberration of light. In particular, he gave hard thought to Michelson and Morley's 1887 ether drift experiment and their disturbing discovery that the motion of the earth had no detectable effect on the propagation of light. Analysis of stellar aberration and other phenomena had provided abundant evidence that the ether of space was stationary and that the earth moved freely through it; yet Michelson and Morley had found no sign of this motion. As Heaviside later observed, the accepted laws of physical optics, and so of electromagnetism, seemed to be faced with "a flat contradiction."[54] FitzGerald, too, had been puzzled by the problem and had spoken about the Michelson-Morley experiment three times at meetings of the Royal Dublin Society early in 1888; Heaviside had told Hertz in February 1889 that the next great problem in electromagnetic theory was "Aberration!"[55] Lodge was in good company when he declared early in 1889 that Michelson and Morley's result posed one of the "outstanding problems" confronting Maxwellian theory.[56]

52. Heaviside to Hertz, 13 July 1889, HH-DM. A remark in Heaviside to Lodge, 4 Aug. 1902, OJL-UCL, suggests that Heaviside and FitzGerald discussed the apparent mass of a moving charge during their 1889 meeting.

53. There is a proof copy of Heaviside 1892, 2: 496–99 [6 Sept. 1889], dated 25 Aug. and corrected in Heaviside's hand, in FG-RDS.

54. Heaviside to Lodge, 13 Nov. 1893, OJL-UCL.

55. Heaviside to Hertz, 14 Feb. 1889, HH-DM; FitzGerald, "Michelson and Morley on the Relative Motion of the Earth and the Luminiferous Aether" (Royal Dublin Society, 18 Jan., 14 Feb., and 18 Apr., 1888), listed (title only) in *Yearbook of Scientific and Learned Societies* (London, 1889).

56. Lodge 1889b: 297–98, first pub. in *Nature 39* (1889): 321; cf. Swenson 1972: 103–5.

This was just the sort of problem at which Lodge excelled: a serious conceptual issue, but one that demanded clear thinking rather than mathematical expertise. It also demanded experimental skill; for Lodge saw that it would be necessary to test the hypothesis that Michelson and Morley's negative result simply showed that moving matter—in this case, the earth itself—carried along the ether immediately surrounding it. Although he thought it unlikely that the ether was actually dragged along in this way, Lodge wanted to explore any possible escape from the contradiction posed by Michelson and Morley's experiment, so as to make the seeming paradox as sharp as possible. He also hoped that a major piece of experimental work would serve to solidify his scientific reputation—for he was acutely aware of having come in second in the race to discover electromagnetic waves. Thus he embarked on an ambitious series of experiments between 1890 and 1893, using an interferometer and an elaborate "whirling machine" to look for evidence of ether drag.[57] He found none, but his work helped keep the problem of motion through the ether prominent in the minds of his British colleagues.

Sometime in the spring of 1889, probably late in March or early in April, FitzGerald paid one of his periodic visits to Lodge in Liverpool, and their conversation quite naturally turned to the Michelson-Morley experiment and the difficulty of reconciling it with the evidence for a stationary free ether. It was then, as the two men sat talking in Lodge's study at 21 Waverley Road, that, in Lodge's words, "the brilliant suggestion . . . flashed on" FitzGerald that the motion of bodies through the ether might cause them to change in size by just the amount needed to account for Michelson and Morley's null result.[58] Lodge later gave at least four accounts of the occasion, ranging from a single sentence in an obituary of FitzGerald he wrote late in 1901 to a full page in his 1931 autobiography, complete with verbatim quotes.[59] As these accounts grow more elaborate, they become less reliable, and it is sometimes difficult to disentangle FitzGerald's own ideas from Lodge's later additions. But it is clear that, while discussing the problem with Lodge, FitzGerald suddenly realized that the motion of Michelson and Morley's interferometer through the ether might have caused it to contract just enough to compensate for the expected shift and so explain their negative result.

The FitzGerald contraction is often presented as a prime example of a purely ad hoc hypothesis, cooked up to explain away a troublesome ex-

57. Lodge 1893a; Hunt 1986.
58. Lodge 1902, in FitzGerald 1902: xxxiv.
59. Ibid.; Lodge 1909: 65–66; Lodge 1913: 25–26; Lodge 1931b: 204–5.

perimental result and only later bolstered with "plausibility arguments," in this case those offered in the mid-1890s by Lorentz and Larmor on the basis of their electron theories.[60] The actual course of FitzGerald's thinking is, of course, inaccessible to us, but it should be stressed that he had in mind the prerequisites for a plausible contraction hypothesis well before his discussion with Lodge precipitated its statement. Although in 1889 he did not yet have a developed electrical theory of matter, FitzGerald thought it quite likely that intermolecular forces were electromagnetic; in any case, he was convinced they depended on the ether just as electromagnetic forces did and so would presumably be affected in the same way by motion. Most important, he had fresh in his mind Heaviside's formula giving just the $\sqrt{1 - v^2/c^2}$ effect needed to account for Michelson and Morley's result. In the spring of 1889, FitzGerald had good reason to think that bodies *should* change in size when they move through the ether; he could in principle have predicted Michelson and Morley's result, given only the plausible additional assumption that intermolecular forces are electromagnetic. Indeed, he apparently told Lodge when the idea first occurred to him (certainly he repeated the remark later) that the contraction *ought* to occur "and this experiment of Michelson's is the only way of detecting it."[61] Lorentz, who followed a similar train of reasoning in 1892, apparently thought much the same way. He wrote to Einstein in 1915 that the contraction hypothesis "would have given much less of an impression of having been invented *ad hoc*" if its close connection to the known laws for moving charges had been brought out more clearly from the first.[62]

FitzGerald usually announced his new ideas at the Royal Dublin Society; it was there, for instance, that he had first published his work on moving charges and on electromagnetic waves in the early 1880s. But his repeated efforts to reform the society led to a confrontation early in 1889 in which his faction was defeated. He resigned as honorary vice-president in February 1889 and had nothing more to do with the society for nearly ten years.[63] With his usual outlet thus blocked, FitzGerald looked for another forum in which to announce his contraction hypothesis. He chose the journal *Science*, apparently because Michelson and Morley were Americans. In a brief letter sent on 2 May and published two weeks later under the title "The Ether and the Earth's Atmosphere,"

60. The account in Eddington 1942: 203 is typical: FitzGerald's hypothesis is described as having been "an *ad hoc* explanation unsupported by anything hitherto known in theory or experiment" until 1895, when Larmor's electron theory "gave it a new status altogether."
61. Lodge 1931b: 205; cf. FitzGerald to Lorentz, 14 Nov. 1894, quoted in Brush 1967.
62. Lorentz to Einstein, Jan. 1915 [draft], quoted in Pais 1982: 167.
63. See Mollan 1981: 210–11.

he wrote that almost the only way to reconcile their result with the abundant evidence for a stationary ether was by the hypothesis that "the length of material bodies changes, according as they are moving through the ether or across it, by an amount depending on the square of the ratio of their velocities to that of light. We know that electric forces are affected by the motion of the electrified bodies relative to the ether, and it seems a not improbable assumption that the molecular forces are affected by the motion and that the size of bodies alters consequently."[64] FitzGerald clearly tied his change-of-size hypothesis to the known way electromagnetic forces were altered by motion, but the letter is brief and elementary, and he did not cite Heaviside's actual formula.

Science was not an especially prominent journal in those days (it had been founded just a few years before and was struggling financially), and it was a poor place for FitzGerald to try to publicize his contraction idea.[65] He rarely saw the journal himself and in fact never saw his own letter in print; nor, it seems, did anyone else of much interest see it at the time. The letter was so obscure it was overlooked when FitzGerald's *Scientific Writings* were collected for publication in 1902—a particular irony since the editor, Joseph Larmor, was FitzGerald's close friend and a great proponent of the contraction hypothesis and of FitzGerald's claim to credit for it.[66]

Once he had sent his letter off to *Science*, FitzGerald did little more to publicize his hypothesis; he often neglected to follow up his ideas very thoroughly, a failing he ascribed to laziness.[67] He mentioned the contraction idea in his lectures at Trinity College Dublin and unsuccessfully urged his former student Thomas Preston to include it in his 1890 textbook *The Theory of Light*.[68] He also told various of his scientific friends about it, though some remained skeptical; R. T. Glazebrook said later that when he and J. J. Thomson heard of the contraction hypothesis during one of FitzGerald's visits to Cambridge as an external examiner, they thought it no more than "the brilliant baseless guess of an Irish

64. FitzGerald 1889, repr. in Brush 1967.
65. Kohlstedt 1980.
66. FitzGerald's letter was first unearthed by Brush in 1967; his discussion corrects that in Bork 1966b. See also Hunt 1988. On Larmor's advocacy of FitzGerald's claims, see Larmor to Lodge, 24 Oct. 1901, OJL-UCL.
67. FitzGerald may also have been put off from following up his initial insight by confusion about the directional variation of the forces between charges moving together, the intricacies of which were not cleared up until 1892 and were not well known for several years after that. See Heaviside 1892, 2: 514n; FitzGerald to Heaviside, 16 and 18 Aug. 1893, OH-IEE; and FitzGerald to Larmor, 30 Mar. and 4 Apr. 1895, JL-RS.
68. FitzGerald to Lorentz, 14 Nov. 1894, quoted in Brush 1967. The FitzGerald contraction is not mentioned in Preston 1890 but is discussed on p. 520 of the 2d ed. (1895) and in the later editions of this widely used text.

genius."[69] Eventually, however, FitzGerald persuaded Lodge to mention the idea in a paper on ether and matter read to the Physical Society in London on 27 May 1892 and published in *Nature* a few weeks later—the first reference to the contraction hypothesis to appear in print since FitzGerald's own letter to *Science* three years before. This was followed late in 1893 by a somewhat fuller (though still very brief) discussion in Lodge's major Royal Society paper "Aberration Problems."[70]

It was via the latter paper that FitzGerald's hypothesis came to the attention of H. A. Lorentz in Holland. Lorentz had hit upon the same idea independently late in 1892 and went on to develop it comprehensively in conjunction with his electron theory. After seeing the brief mention of the contraction hypothesis in Lodge's "Aberration Problems," he wrote FitzGerald in November 1894 to ask if he had ever published anything on the subject. FitzGerald replied that he recalled writing the letter to *Science* but doubted it had been published. He was glad to hear that Lorentz was working on the problem and seemed content to leave its further development to others.[71] Lorentz inserted a footnote in his major work on electron theory, his *Versuch* of 1895, mentioning that FitzGerald had also proposed the contraction hypothesis, and it is largely because of this reference that the effect has become known as the "FitzGerald-Lorentz contraction."[72]

What does this account of the origin of FitzGerald's contraction hypothesis tell us, not just about the evolution of an important physical idea but about the operation of the Maxwellian group? First, it shows how important personal interactions and seemingly extraneous circumstances can be in the birth and dissemination of scientific ideas. Had it not been for Heaviside's campaign for recognition, for instance, or FitzGerald's tiff with the Royal Dublin Society, the story of the FitzGerald contraction might have been very different.

At a deeper level, the origin of the FitzGerald contraction illustrates in some detail the different roles the various Maxwellians played, not just in this episode but in their other joint work as well. Thus we see

69. Glazebrook 1928; Glazebrook thought FitzGerald's visit was probably in June 1893, but a check of Cambridge records by Andrew Warwick shows that FitzGerald did not serve as an external examiner until June 1894. For more on the early days of the contraction hypothesis, see Warwick 1989: 28–30; Warwick emphasizes that most other British physicists did not find FitzGerald's hypothesis persuasive until it had been integrated into an electronic theory of matter.
70. Lodge 1892c; Lodge 1893a: 749–50.
71. Lorentz's letter of 10 Nov. 1894 and FitzGerald's reply of 14 Nov. are reproduced in Brush 1967.
72. Lorentz 1895: 122n; cf. FitzGerald to Larmor, 30 Mar. 1895, JL-RS.

FitzGerald, both at the Bath meeting and with Lodge in Liverpool, as a man who excelled in discussions, whose strength lay not in quietly pondering scientific questions in his study or laboratory but in talking them out with his friends and colleagues. It was in such discussions that FitzGerald's best ideas and most searching criticisms were usually formulated. His talents were complemented by those of Lodge, who served as an excellent sounding board for FitzGerald's ideas and a very effective expositor of their significance. It was Lodge's clear and forceful account of the "murder of ψ" debate in the *Electrician* that first brought the underlying issues to the attention of those not at Bath, notably Heaviside; and it was mainly through Lodge's lectures and publications, particularly *Modern Views,* that Maxwellian ideas began to reach a wide audience in the 1880s and 1890s. Lodge was also a skillful and energetic experimenter, as his earlier work with waves on wires had shown, and his whirling-machine experiments played an important part in tightening the logical noose that made the contraction hypothesis seem inescapable. Lodge was in no sense the originator of the contraction hypothesis, but he played a catalytic role in its formulation and later dissemination, just as he did in the formulation and dissemination of the leading principles of the Maxwellian synthesis as a whole.

Heaviside appears in the FitzGerald contraction episode in two of his main roles in the Maxwellian group: as chief mathematical theorist and as defender of the Maxwellian faith. "I preach the gospel according to my interpretation of Maxwell," he declared in one of his early papers on the moving-charge problem, and it was he who rose to William Thomson's challenge and showed how the propagation of electromagnetic effects could be treated on purely Maxwellian grounds.[73] Heaviside had done more than anyone else in the 1880s to refine and extend the mathematical structure of Maxwell's theory and to bring out its full implications. FitzGerald and Lodge often turned to him for solutions to difficult mathematical problems, particularly on the generation and propagation of electromagnetic waves, and it was no accident that it was Heaviside who first found the complete formula for the field around a moving charge, the key to a plausible contraction hypothesis.

Once this formula had been found, the thread was taken up by Lodge and FitzGerald, and two more aspects of their roles came into play. Lodge often described himself as a "brooder"; he would spend weeks turning scientific problems over and over in his mind, looking for a satisfactory solution.[74] In the spring of 1889 he was pondering his favor-

73. Heaviside 1892, 2: 492 [1888].
74. Lodge 1931b: 110–11, 343.

ite topic, the ether, and the implications for it of Michelson and Morley's experiment. When he presented the different aspects of the problem to FitzGerald, his friend responded, as so often before, with a flash of insight, or (depending on one's point of view) a rather wild suggestion. In a letter to Lodge, Heaviside once referred to FitzGerald as "our friend of brilliant ideas"; FitzGerald was always full of ideas, some of them indeed brilliant, nearly all of them stimulating.[75] In one of his first letters to Heaviside, FitzGerald had said that "as I am not in the very least sensitive to having made mistakes I rush out with all sorts of crude notions in hopes that they may set others thinking and lead to some advance."[76] The presence of such an "idea man" can be crucial to the work of a group of scientists; while he is not likely to make solid or profound advances on his own, in combination with the right mix of critical and hard-headed colleagues, he can stimulate very valuable work.

Shortly after FitzGerald's death, Heaviside paid tribute to his friend's gifts while commenting insightfully on their drawbacks. "He had, undoubtedly, the quickest and most original brain of anybody," Heaviside wrote to John Perry. "That was a great distinction; but it was, I think, a misfortune as regards his scientific fame. He saw too many openings. His brain was too fertile and inventive. I think it would have been better for him if he had been a little stupid—I mean not so quick and versatile, but more plodding. He would have been better appreciated, save by a few."[77] Heaviside and Lodge were among those who could appreciate FitzGerald as he was, and in the origin of the contraction hypothesis, as in their other work in the 1880s and 1890s, it was the combination of their complementary talents that enabled them to bring out the full meaning of Maxwell's theory.

What Is Maxwell's Theory?

The formulation of the new Maxwellian synthesis raised an important question: Was this still "Maxwell's theory," or was it now Heaviside's, or Hertz's, or FitzGerald's theory? In what sense could such books as Lodge's *Modern Views* or Heaviside's *Electrical Papers* be considered expositions of Maxwell's ideas? Just what did the phrase "Maxwell's theory" now mean? The Maxwellians raised these questions quite explicitly in the early 1890s, prompted partly by the spread of Maxwell's theory to

75. Heaviside to Lodge, 19 Jan. 1901, OJL-UCL.
76. FitzGerald to Heaviside, 4 Feb. 1889, OH-IEE.
77. Heaviside to Perry, [Feb. 1901], quoted in Larmor 1901a, in FitzGerald 1902: xxvi; cf. Larmor to Heaviside, 6 Mar. 1901, OH-IEE.

the Continent and the consequent formulation of variant interpretations. Their answers provide a valuable insight into the nature of the Maxwellian synthesis; in particular, they point up the central role in that synthesis of ideas about the localization and flux of electromagnetic energy.

The most famous answer to the question What is Maxwell's theory? was that Hertz gave in 1892 in the introduction to his *Electric Waves:* "I know of no shorter or more definite answer than the following:—Maxwell's theory is Maxwell's system of equations."[78] Later writers have made much of this remark, seeing in it the key to Hertz's rather austere attitude toward Maxwell's theory and toward physical theorizing in general.[79] But while it is true that Hertz's remark reflected his emphasis on clarity and logical consistency in the statement of theories, he also had a more immediate reason for defining Maxwell's theory in this way. Hertz had come to Maxwell's theory through Helmholtz's generalized system of electrodynamics and had arrived at a set of equations from which Maxwell's key concept of dielectric displacement in free space (as distinct from electric force) had been eliminated. He wanted to justify regarding both Helmholtz's equations and his own as valid expressions of "Maxwell's theory." To do this, he had to separate the mathematical content of Maxwell's theory, to which he believed his own and Helmholtz's equations to be formally equivalent, from its physical basis, with which Hertz's and Helmholtz's theories were in conflict.

Helmholtz had continued to base his theory on the image of charges and currents exerting forces on each other directly from a distance. By imagining the intervening space to be filled with a polarizable medium and then analyzing how the original forces would be modified by the induced polarizations, he was able to derive equations that bore a strong formal resemblance to Maxwell's. But the physical basis of Helmholtz's theory in fact evaporated when it was taken to the "Maxwell" limit; the direct forces between the charges and currents were completely swamped by the induced polarizations, and the physical nature of "charge" and "current" was hopelessly obscured.[80] Heaviside's assessment was harsh but perceptive: "Helmholtz's theory seemed to me as if he had read all of Maxwell at once, then gone to bed and had a bad dream about it, and then put it down on paper independently; his theory being Maxwell's run mad."[81] Helmholtz had failed to grasp, or at least

78. Hertz 1893: 21.
79. Heimann 1971: 156–57; Knudsen 1985: 176–78.
80. Hertz 1893: 23–26; Buchwald 1985a: 177–86.
81. Heaviside to Lodge, 15 June 1892, OJL-UCL; see also Heaviside notebook 7, p. 69, OH-IEE, where he compared Helmholtz's theory to "Maxwell gone mad, or drunk."

to accept, Maxwell's idea that "charge" was simply the result of strain in the ether, and his attempt to fit Maxwell's theory into an action-at-a-distance framework inevitably ran aground.

In the introduction to his *Electric Waves,* Hertz testified to the difficulty those trained in Continental electrodynamics had in grasping Maxwell's ideas, especially in the ambiguous and even contradictory way they were presented in his *Treatise:* "Many a man has thrown himself with zeal into the study of Maxwell's work, and, even when he has not stumbled upon unwonted mathematical difficulties, has nevertheless been compelled to abandon the hope of forming for himself an altogether consistent conception of Maxwell's ideas."[82] Hertz had fared no better himself, he said, and had finally decided that it was best simply to leave aside many of Maxwell's physical notions, such as "displacement," and to focus solely on his equations. It was these equations, whether in Maxwell's form, or Helmholtz's, or Hertz's, that represented "the undying part of Maxwell's work," according to Hertz; they, "and not Maxwell's peculiar conceptions or methods, would I designate as 'Maxwell's Theory.' " In this way, Hertz was able to preserve his allegiance to Helmholtz without continuing to endorse his physical conception of the electromagnetic field. It was only in the limited sense of being mathematically equivalent to Maxwell's equations and so embracing the same possible experimental results that Hertz's two papers of 1890 could be regarded as expressions of "Maxwell's theory." "In no sense can they claim to be a precise rendering of Maxwell's ideas," Hertz admitted. "On the contrary, it is doubtful whether Maxwell, were he alive, would acknowledge them as representing his own views in all respects."[83]

In looking at Maxwell's theory as simply a set of equations, Hertz broke ranks with his British colleagues. An October 1893 review in *Nature,* unsigned but almost certainly by FitzGerald, gave the German edition of Hertz's collected electrical papers high praise but noted, with concern:

> Prof. Hertz . . . seems content to look upon Maxwell's theory as the series of Maxwell's equations. This is hardly fair. Maxwell has done much more than produce a series of equations that represent electromagnetic actions. Weber and Clausius went very close to that without revolutionising our ideas as to the nature of these actions. Any exposition of Maxwell's theory

82. Hertz 1893: 20.
83. Hertz 1893: 21; see also Hertz to Helmholtz, 24 Feb. 1892, in Hertz 1977: 321, in which Hertz asks for permission to dedicate his volume on electric waves to Helmholtz and says that his "Introduction" will make it clear that his own work "derives not merely from the direct study of Maxwell's works, as I am constantly told, but rather essentially from the study of the works of Your Excellency [Helmholtz]."

which does not clearly put before the reader that energy is stored in the ether by stresses working on strains, is a very incomplete representation of Maxwell's theory.[84]

These stresses and strains and this energy in the ether were, to the British Maxwellians, the heart of Maxwell's theory. They thus objected not just to Hertz's reduction of Maxwell's theory to a bare set of equations but more particularly to his abandonment of the key physical concept of dielectric displacement in the free ether. "As regards Maxwell's theory," FitzGerald told Heaviside late in 1893,

> I am afraid I consider it of the essence of that theory to distinguish between Force and Displacement. It is generally almost ignored and the ether in consequence is also ignored everything resting on symbols to the destruction of a chance of explaining the structure of the ether and also to the confusion of learners who, I think require some geometrical and dynamical analogues to hang their arguments on. I would call any reasonable modification of this fundamental idea, Maxwell's Theory, though perhaps this germ is really Faraday's.[85]

Both force and displacement, or stress and strain, were needed to account dynamically for the storage of energy in the ether. Hertz had defined the electric and magnetic constants as unity in free space, thus eliminating displacement as a primary quantity. While this somewhat simplified the mathematical structure of Maxwell's theory, it was fatal to its dynamical basis. Heaviside raised this point with Hertz in 1890 after reading the first of his papers "On the Fundamental Equations of Electrodynamics." It was "very interesting to see the points of difference of view," Heaviside said, especially about force and displacement and the constants connecting them, but he was not happy with Hertz's conclusions. "Can you conceive of a medium for el[ectro]-mag[netic] disturbances," he asked, "which has not at least *two* physical constants, *analogous* to density and elasticity? If not, is it not well to *explicitly symbolize them,* leaving to the future their true interpretation?"[86] Heaviside was calling on Hertz not to sacrifice, for the sake of a small and perhaps illusory mathematical simplification, the dynamical ether, filled with stresses,

---

84. [FitzGerald] 1893. Internal evidence of FitzGerald's authorship includes remarks that parallel those he made in "M. Poincaré and Maxwell," FitzGerald 1902: 284–87 [1892]. There is a copy of Hertz 1892 (German ed.) in FG-TCD Physics; its pages are uncut except for the "Introduction," but FitzGerald was already familiar with most of the papers reprinted in the volume and had recently read part of a draft of D. E. Jones's translation; see Jones to Hertz, 17 Oct. 1893, quoted in O'Hara and Pricha 1987: 104.
85. FitzGerald to Heaviside, 22 Dec. 1893, OH-IEE.
86. Heaviside to Hertz, 8 Dec. 1890, HH-DM.

strains, and stored energy, on which Maxwell had built his theory. The British Maxwellians took this dynamical ether much more seriously than did their Continental counterparts, insisting that even on those points where Maxwell's theory required clarification and correction, this could and should be done without reducing the theory to a mere set of equations.

The important modifications Heaviside had made to Maxwell's theory, including his abandonment of the potentials in favor of his "duplex" force equations, were not, he said, meant as substantive changes or new departures but were directed solely at bringing out the leading points of the theory more clearly than had Maxwell himself. Heaviside drew a careful distinction, as did FitzGerald, between Maxwell's *Treatise* and Maxwell's *theory;* the book, both said, gave only an imperfect account of the real nature of the theory.[87] Heaviside made this point in December 1893 in the preface to the first volume of his *Electromagnetic Theory.* In addressing the question What is Maxwell's theory? he said:

> The first approximation to the answer is to say, There is Maxwell's book as he wrote it; there is his text, and there are his equations: together they make his theory. But when we come to examine it closely, we find that this answer is unsatisfactory. To begin with, it is sufficient to refer to papers by physicists, written say during the twelve years following the first publication of Maxwell's treatise, to see that there may be much difference of opinion as to what his theory is. It may be, and has been, differently interpreted by different men, which is a sign that it is not set forth in a perfectly clear and unmistakeable form. There are many obscurities and some inconsistencies. Speaking for myself, it was only by changing its form of presentation that I was able to see it clearly, and so as to avoid the inconsistencies. Now there is no finality in a growing science. It is, therefore, impossible to adhere strictly to Maxwell's theory as he gave it to the world, if only on account of its inconvenient form. But it is clearly not admissible to make arbitrary changes in it and still call it his. He might have repudiated them utterly. But if we have good reason to believe that the theory as stated in his treatise does require modification to make it self-consistent, and to believe that he would have admitted the necessity of the change when pointed out to him, then I think the resulting modified theory may well be called Maxwell's.[88]

Heaviside was critical of those who hewed too closely to Maxwell "as he was wrote." He complained to Hertz early in 1889 that "most writers

87. See FitzGerald to Larmor, 24 Mar. 1894, JL-RS. Hertz also drew this distinction but felt less certain than the British of Maxwell's true meaning; see Hertz 1893: 20.

88. Heaviside 1893–1912, vol. 1, "Preface" (unpaginated); it was in response to an advance copy of this preface that FitzGerald made the remarks quoted above.

here follow Maxwell slavishly, and repeat his faults and errors, even though they may sometimes disobey the spirit of his treatise in following the letter."[89] This tendency to follow Maxwell to the letter was strongest at Cambridge, where his *Treatise* was long accorded the status of "a species of sacred writing," according to one old Cambridge man, and was treated more as a source of mathematical puzzles than as a physical theory subject to clarification and amendment.[90]

J. J. Thomson, who in 1884 had succeeded to the Cavendish Professorship first held by Maxwell and in 1892 edited the third edition of the *Treatise,* was especially "fettered by his connection with Maxwell," in Heaviside's words, and in fact did very little to update the presentation of Maxwell's theory even after Heaviside, FitzGerald, and Hertz had pointed the way.[91] Like most Cambridge mathematicians, Thomson had been intensively trained in the use of potential and Lagrangian functions and had come to regard the extension of these methods as synonymous with the progress of mathematical physics. He and his Cambridge colleagues quite naturally resisted any attempt to replace this approach with one that focused more closely on the actual forces and flow of energy in the field—especially when that attempt was led by someone as far outside the Cambridge orbit as Oliver Heaviside.[92]

Heaviside's work was rooted in concrete technological problems, and though he made himself a master of the most arcane mathematical methods, he always insisted one should "keep as near to the physics of the matter as one can, and not be deluded by mere mathematical functions."[93] There was much in Heaviside's approach that was foreign to the mentality of late-Victorian Cambridge, where technology had a very low status (smelling, as it did, of "trade") and even experiment was slow to

---

89. Heaviside to Hertz, 14 Feb. 1889, HH-DM.

90. Charles Chree, "Mathematical Aspects of Electricity and Magnetism," *Nature 78* (1908): 537. Like many sacred writings, Maxwell's *Treatise* was not much read by undergraduates; see Goldberg 1970: 124 for evidence that it was not part of the usual Cambridge curriculum until after 1900.

91. Heaviside to W. G. Bond, 28 Mar. 1896, filed among Heaviside's letters to Lodge, OJL-UCL. Thomson mentioned Heaviside's form of Maxwell's equations in his notes to Maxwell 1892, art. 603, saying it had "the merit of simplicity," but deprecated it as less general than the vector potential approach and did not use it in his own work.

92. Buchwald 1985c: 304 also draws a sharp contrast between Heaviside and the Cambridge school, but he depicts Heaviside as a "Maxwellian apostate" who was intellectually isolated from "the Maxwellian mainstream." This characterization reflects an important difference in how Buchwald and I use the term *Maxwellian:* where Buchwald identifies it with the Cambridge school and J. J. Thomson's Lagrangian approach, I contend that the main line of Maxwellian development lay outside Cambridge and centered on Heaviside, FitzGerald, Lodge, and Hertz, all of whom moved away from the Lagrangian methods Maxwell himself used. For a supporting characterization of the Maxwellian group, see Warwick 1989: 12.

93. Heaviside to Hertz, 13 July 1889, HH-DM.

penetrate the dominant tradition of mathematical physics. Cambridge men dominated British physics throughout the nineteenth century, but it was no accident that when it came to reforming the interpretation of Maxwell's theory and confirming it experimentally, the leading figures—Heaviside, FitzGerald, Lodge, and Hertz—all came from outside Cambridge. When Heaviside told Hertz in July 1889 that "the little progress made for many years in developing Maxwell was due to a too slavish attempt to follow him with his ψ and $\mathbf{A}$," he might have added that the real roots of this seeming "slavishness" lay in the deeper commitments of Cambridge mathematicians.[94]

J. J. Thomson's own early career offers several examples of the pitfalls of following Maxwell too closely. The major flaws in Thomson's famous 1881 paper on moving charges arose because he took certain statements and equations in the *Treatise* literally even when they conflicted with such basic principles of Maxwell's theory as the requirement that all currents form closed circuits. He had fallen into similar difficulties in an 1880 paper on the propagation of light in moving media, again by following the letter of the *Treatise* rather than its spirit. By applying Maxwell's equations, Thomson found that a dielectric medium moving with velocity $\mathbf{v}$ would partially drag light waves along within it, adding $\frac{1}{2}\mathbf{v}$ to their speed. This contradicted Fresnel's well-known drag coefficient of $1 - 1/n^2$, where n is the index of refraction of the moving medium. But water, the only medium for which Fresnel's drag hypothesis had been experimentally tested, has an index of refraction such that, within experimental error, $1 - 1/n^2 = \frac{1}{2}$, so Thomson could point to these experiments as also confirming his own (and so, he thought, Maxwell's) theory of the propagation of light in moving media.[95]

Despite its agreement with experiment and its apparent basis in Maxwell's theory, Thomson's $\frac{1}{2}\mathbf{v}$ drag theory attracted little support; for it seemed to violate the basic physical principle of the relativity of motion. When Lodge was beginning his whirling-machine experiments on ether motion in 1891, he thought Thomson's result might be relevant and asked FitzGerald for his opinion of it. FitzGerald pointed out that since Thomson's $\frac{1}{2}\mathbf{v}$ drag was independent of the nature of the moving medium, it implied that a moving portion of even a very tenuous medium— even of the ether itself—would pull light waves along at half its own velocity.[96] But its velocity relative to what? Thomson's hypothesis clearly required an absolute frame to which motion of the ether itself could be

94. Ibid.; on the disciplinary context of Cambridge mathematics, see Hunt 1991.

95. J. J. Thomson 1880: 291. The water experiment had first been done by Hippolyte Fizeau in 1851.

96. Lodge to FitzGerald, [Apr. 1891], FG-RDS, and FitzGerald to Lodge, [late Apr. 1891] and 27 Apr. 1891, OJL-UCL; cf. Lodge 1893a: 732–33.

referred—a notion FitzGerald and other classical physicists found un-
acceptable. Moreover, since Maxwell's theory treated any homogeneous
dielectric as a simple medium—essentially a portion of the ether with
different values for the electric and magnetic constants—which carried
all its properties along with it, there *must* be full superposition of propagation velocities in a moving body; thus the theoretical drag coefficient
had to be 1, not $\frac{1}{2}$ or $1 - 1/n^2$. FitzGerald and Lodge thus concluded on
general principles that Thomson's result must be based on a mistaken
interpretation of Maxwell, though they were not then able to track down
the actual source of the error.

Heaviside had solved the moving-medium problem for himself in
1889 and had shown that Maxwell's theory implied a full **v** drag.[97] When
Lodge brought Thomson's earlier work to his attention in October 1893,
Heaviside's response revealed much about his attitude toward Maxwell's
theory: "It is no go," he said of Thomson's result. "All very well in 1880,
when Maxwell wasn't understood by *anybody*, not even by himself! . . . J.
J. T. is pure and unadulterated Maxwell *as he was wrote*," Heaviside observed, and this fidelity to the letter of the *Treatise* was the source of
Thomson's errors. "It is not generally recognized," Heaviside said, "that
Maxwell the man did not publish a consistent and complete scheme for
moving media. Why he missed it was because of his vector and scalar
potentials. It is a hopelessly obscure business in terms of them, and I
doubt whether anyone could do it, unless he knew results by other
ways."[98] Maxwell had introduced what Heaviside called a "motional
electric force" $\mathbf{e} = \mathbf{v} \times \mathbf{B}$ (where **v** is the velocity of the medium and **B** is
the magnetic induction) to account for the induction of currents by mo-
tion in a magnetic field, and it was from this—or rather from its equivalent in terms of the potentials—that Thomson had derived his $\frac{1}{2}\mathbf{v}$ drag.
But Heaviside, using his "duplex" set of equations, had shown that, to
preserve symmetry, there must also be a motional magnetic force $\mathbf{h} = \mathbf{D} \times \mathbf{v}$ (where **D** is the dielectric displacement) for motion in an electric
field. Each of the motional terms contributes a drag of $\frac{1}{2}\mathbf{v}$; together they
yield a full **v** drag so that the velocity through the medium and the velocity of the medium are added together, as on general principles they
must be.

This motional magnetic force was Heaviside's invention; it is not in
Maxwell's *Treatise* at all. As Heaviside told Lodge, "You may, if you like,
and with some sense, say it is not Maxwell, but is Heaviside."[99] But Heav-

97. Heaviside 1892, 2: 519–21 [1889]; see also Whittaker 1951: 330–31.
98. Heaviside to Lodge, 30 Oct. 1893, OJL-UCL.
99. Ibid. Heaviside's "motional forces" are discussed more fully in Buchwald 1985c:
310–12.

iside contended that we should "regard the modified form as truly expressive of Maxwell's theory rightly interpreted."[100] It represented "the real and true 'Maxwell' as Maxwell would have done it had he not been humbugged by his vector and scalar potentials"; on this, as on many other issues, Heaviside believed that Maxwell "would have accepted the change at once when pointed out to him, and have adopted it in his own work."[101] Heaviside claimed he was simply expounding Maxwell's teachings but doing so more clearly, consistently, and thoroughly than Maxwell had himself. He did not regard the fact that the experimentally observed drag coefficient was less than one as proof that Maxwell's basic field theory was wrong but simply as evidence that it would have to be supplemented by a more detailed theory of material media—a theory of their microscopic structure, like that the electron theory later provided.

Heaviside believed that one should neither follow Maxwell's *Treatise* to the letter nor throw it all out except for a few equations. His aim was, rather, to extract from "that mine of wealth 'Maxwell'" a few leading ideas, mainly those concerning the stress, strain, and energy in the ether, and to construct on that basis a clear and consistent theory that remained true to Maxwell's principles. This revised and purified version constituted "the true Maxwellian theory" in Heaviside's eyes and in those of FitzGerald, Lodge, and their followers; and it was the one they sought to promote in the late 1880s and 1890s.[102]

This process of revision tended to put a certain distance between "Maxwell the man" and the theory that went by his name; indeed, in 1895, Heaviside went so far as to say that "Maxwell was only $\frac{1}{2}$ a Maxwellian."[103] Heaviside's own definition of "Maxwellian" was simple: "By a Maxwellian," he had told FitzGerald in 1889, "I mean one who follows Maxwell as interpreted by O. H."[104] This was said partly in jest, but it contained considerable truth. With his emphasis on the primacy of the electric and magnetic forces and on the energy stored in the field, Heaviside had expressed the characteristic ideas of the new synthesis much more clearly and consistently than had Maxwell himself, and he had laid down the lines the other Maxwellians were subsequently to follow.

Lodge played a key role in bringing these ideas to a broader scientific public, and though he generally expressed them more through models

100. Heaviside to Lodge, 28 Oct. 1893, OJL-UCL.

101. Ibid., 30 Oct.

102. Heaviside 1892, 2: 393 [1888], and Heaviside 1893–1912, vol. 1, "Preface" (unpaginated).

103. Heaviside to FitzGerald, [Mar. 1895], FG-RDS; internal evidence places this undated fragment between FitzGerald's letters to Heaviside of 8 and 15 Mar. 1895, OH-IEE.

104. Heaviside to FitzGerald, 30 Jan. 1889, FG-RDS.

than through equations, many of the chief points in his *Modern Views,* especially concerning the stress, strain, and energy in the ether, closely paralleled the themes Heaviside had emphasized in his own expositions of Maxwell's theory. When Heaviside called *Modern Views* the first "popularisation of Maxwellian views," he was, in light of the meaning he gave the term *Maxwellian,* giving a strong indication of the accord he saw between Lodge's views and his own.[105]

The affinity between Heaviside's views and FitzGerald's was even stronger. Heaviside had high praise for the work FitzGerald had done in the early 1880s on "the possibilities of Maxwell's theory, then considerably undeveloped and little understood" and declared that "his way of looking at things was more like my own than anybody's."[106] Early in his career, FitzGerald had followed Maxwell's own methods fairly closely, as when he used the potentials and the principle of least action in his 1878 paper on the reflection and refraction of light. But after about 1885, led by his wheel-and-band model and associated ideas about the flow of energy, he moved toward a position almost identical to Heaviside's, in which the real core of Maxwell's theory was taken to be the electric and magnetic forces and fluxes and the energy localized in the field. This interpretation comes across especially clearly in FitzGerald's 1893 review of Heaviside's *Electrical Papers.* Heaviside's "great advance in method" was due, FitzGerald said, "to his consistently following out the principles that Maxwell laid down in accordance with Faraday's ideas as to the nature of electromagnetic actions"—more consistently, in fact, than had Maxwell himself. Although "Maxwell had clearly pointed out that the state of the ether at each point depended upon the electric and magnetic forces there, and upon the electric and magnetic displacements they produced," he had seemingly lost the thread of this central idea among the "maze of symbols" for the electric, magnetic, and vector potentials that filled his *Treatise.* Heaviside had now cleared away this "*débris,*" FitzGerald said, and so at last laid bare the true meaning of Maxwell's theory.[107]

FitzGerald ended his review with a call for "some author with a genius for popular exposition" to write a "New Electromagnetism," a textbook "with electric pressure and the displacement it produces as the source of the electric energy per unit volume in the ether, and with magnetic momentum and its accompanying flow as the source of the magnetic energy per unit volume in the ether" as its leading ideas. "Such a work," he said,

105. Heaviside to Lodge, 20 July 1889, OJL-UCL.
106. Heaviside to Perry, [Feb. 1901], quoted in Larmor 1901a, in FitzGerald 1902: xxvi.
107. FitzGerald 1902: 294 [1893].

"would be electromagnetism up to date, and its possibility would be largely due to Oliver Heaviside."[108]

The forces, fluxes, and especially the flow of energy in the field were the leading themes in Heaviside's writings—but not in his alone. They were in fact the hallmarks of the Maxwellian synthesis as a whole. When FitzGerald told Heaviside in 1892 that an emphasis on the energy flux principle "seems to me to distinguish the Physicist from the Old Tory who sticks to the old ideas on the one hand and the Mathematician . . . on the other," he might almost have said that it distinguished the true Maxwellians from the rest.[109] It is significant that Hertz, who was so close to the British Maxwellians in most other ways, was rather skeptical about the idea of the localization and flux of energy through the ether.[110] Having denied himself the unifying physical conception of energy flux in a dynamical ether, Hertz was left with little more of Maxwell's theory than a bare set of equations.

For the British Maxwellians, Maxwell's theory was something far more vivid. It was the theory—or at least a step toward the theory—of a dynamical ether, the seat of stresses and strains as real as those in a piece of machinery or in one's own muscles and sinews. It was a theory from which the notion of direct action at a distance had been wholly banished and in which the actions exerted at a point were referred solely to the physical state of the medium—the forces, fluxes, and energy—at that point. It was with this interpretation, and in the equations in which these fundamental ideas were implicit, that "Maxwell's theory" passed into general circulation in Britain in the 1890s.

The transition to this new interpretation can be traced very strikingly in the successive editions of *The Theory of Light* by FitzGerald's protege Thomas Preston. An excellent textbook that was reportedly used "by the great majority of advanced students of physics, both in Great Britain and America," it included an important chapter on the electromagnetic theory of light.[111] In the first edition, published in 1890, Preston followed Maxwell "as he was wrote," deriving the electromagnetic wave equation from terms involving the vector potential and giving little prominence to the electric and magnetic forces or the energy in the field. But by the time the more widely read second edition appeared in 1895, the new regime was in place: Preston completely rewrote the section on Maxwell's theory, abandoning the vector potential in favor of

108. Ibid., p. 300.
109. FitzGerald to Heaviside, 26 Sept. 1892, OH-IEE.
110. Hertz 1893: 220 [1890].
111. "Thomas Preston" [obituary], *Physical Review 11* (1900): 188.

Heaviside's "duplex" equations and focusing firmly on the electric and magnetic forces and the energy localized in the field.[112] Similar treatments appeared in other textbooks in the mid-1890s.[113] By the end of the decade, the Maxwellian form of the equations had become standard—though important new trends were already beginning to show themselves.

The "Maxwellian heyday," roughly 1888 to 1894, was the most important period in the spread of Maxwell's theory. It brought to fruition the work that Maxwell had begun and that the Maxwellians had deepened, revised, and in many ways transformed. Maxwell's theory—for despite all the changes it was still firmly rooted in his ideas—emerged in the late 1880s and early 1890s as a solid, consistent, and well-confirmed theory. Electromagnetism was far from being a completed science; the advent of electron theory in the later 1890s and the torrent of new experimental and theoretical discoveries that followed in the twentieth century attest to that. But by the mid-1890s, the Maxwellians had jointly and solidly laid the foundations of the theory of the electromagnetic field, and it was their revised and clarified form of Maxwell's theory that provided the basis on which the new advances could be built.

112. Compare Preston 1890: 444–47, with the 2d ed. (1895), pp. 549–52, which was repeated unchanged in later editions. Preston followed Heaviside quite closely in the second and later editions, even citing his idea of magnetic conductivity, but did not introduce his vector notation.

113. One of the best and most influential was Föppl 1894, which drew heavily on Heaviside's work; see the praise of it in Heaviside 1893–1912, 3: 504 [1897]. Einstein was among the many German physicists who learned Maxwell's theory from Föppl's book; see Miller 1981: 151.

CHAPTER NINE

The Advent of the Electron

Oliver Lodge ended his *Modern Views of Electricity* early in 1889 with a proud boast. Hertz's experiments had "utterly and completely verified" Maxwell's electromagnetic theory of light, he said, and had made it one of the strongest in all of physics. Casting this triumph in very up-to-date terms, he declared, "The whole domain of Optics is now annexed to Electricity, which has thus become an imperial science."[1] But like many imperial annexations, this one proved hard to digest. Maxwell's theory gave an excellent account of purely field phenomena, but it said little that was definite and much that was incorrect about the interaction of fields with matter, particularly at optical frequencies. As the Maxwellians sought to secure their hold over the newly claimed optical and material realm, they were driven to extend and modify their theory in important ways and to introduce a new constituent into their universe: the electron.

As Heaviside and FitzGerald often emphasized, Maxwell's was a theory of *one* medium; it drew no real distinction between ether and matter and treated material bodies simply as portions of the ether in which the electric and magnetic constants (the conductivity, permittivity, and permeability) took on different values.[2] This essentially macroscopic approach enabled the theory to account for a wide range of phenomena with a minimum of hypothesis, but the Maxwellians were well aware that it did so only by slurring over the details of the underlying processes.

1. Lodge 1889b: 307; orig. in *Nature 39* (1889): 322.
2. Heaviside to FitzGerald, 25 Dec. 1893, FG-RDS; cf. FitzGerald 1902: 511–12 [1900].

Maxwell's theory was only "a sort of skeleton-framework," Heaviside told Hertz in 1889; it would not be complete until it was made a "molecular theory as well."[3]

The need to go beyond the old macroscopic approach was especially pressing, and the task especially difficult, for phenomena involving electrical conduction. The existence of conductivity was crucial to Maxwellian theory; without it, there could be no continuous currents, nor even any charges, since it was only by the conductive decay of electric strain that the discontinuities of displacement that constituted "charge" could appear. Yet neither Maxwell nor his successors were able to explain *how* conducting matter broke down electrical strain and dissipated it as heat; they could give only the laws, not the mechanism, of electrical conduction. Their simple single-medium theory ran into even more problems, including direct conflict with experimental results, when it was applied to optical phenomena. It could not explain the low opacity of metal foils, or dispersion, or the partial dragging of light waves by moving media, or a number of puzzling magneto-optic effects. All these phenomena pointed toward some underlying microscopic structure of matter, and perhaps of the ether as well. By the mid-1890s the explication of this substructure had emerged as one of the principal tasks of electrical physics.

All the Maxwellians contributed to the effort to find a molecular basis for electromagnetism, but it was a new recruit to their ranks, Joseph Larmor, who took the lead in this work in Britain. In the fall of 1893, Larmor began to elaborate a new and extremely ambitious dynamical theory of ether and matter, and after several false starts, and with considerable prodding from FitzGerald, he found that many of the longstanding problems in linking fields to matter could be resolved by postulating the existence of "electrons," conceived as mobile singularities in the electromagnetic ether. Lorentz had begun working on similar lines a short time before, and after 1895 their two projects largely merged. The advent of the electron in some ways marked a sharp break with previous Maxwellian theory; in others, it was perhaps its natural culmination. Either way, it clearly opened a new stage in the development of electrical theory.

Joseph Larmor and the Rotational Ether

Joseph Larmor was an Ulsterman, born in County Antrim in 1857 and educated at the Royal Belfast Academical Institution and Queen's Col-

---

3. Heaviside to Hertz, 14 Aug. and 13 July 1889, HH-DM; cf. FitzGerald to Larmor, 29 Mar. 1894, JL-RS.

Joseph Larmor, about 1912. From *Mathematical Gazette*, 1912

"As regards Larmor, if Maxwell was only ½ a Maxwellian, L. is only ¼ or less."
—Oliver Heaviside to G. F. FitzGerald, March 1895

lege, Belfast. His mathematical talents then carried him to Cambridge, where he finished as senior wrangler in 1880. (J. J. Thomson came second.) After a stint teaching natural philosophy at Queen's College, Galway, he returned to Cambridge in 1885 to take up a lectureship in mathematics at his old college, St. John's. A reserved and somewhat diffident man, Larmor soon fell into the routine of a Cambridge mathematics don. He taught and practiced "mixed mathematics" of the kind long traditional at Cambridge and was noted for what Arthur Eddington later called his "intense, almost mystical, devotion to the principle of least action," which he regarded as the proper foundation of all physical theory.[4] He published a few papers on electrical topics while at Galway but turned mainly to geometrical optics and analytical dynamics after his return to Cambridge and did not take up electromagnetism seriously again until after 1890. He knew FitzGerald through their mutual friends in Ireland but had little other contact with the Maxwellian group in the 1880s.

This began to change in 1893, when Larmor's interest in optics led him to examine the action of magnetism on light. He presented a long paper on the subject to the British Association in September; in it he surveyed the various theories of light and identified the main problems in the treatment of magneto-optic phenomena.[5] He gave particular attention to MacCullagh's rotationally elastic ether and to the electromagnetic version of it FitzGerald had devised in 1878, correcting a subtle error in FitzGerald's boundary conditions and extending the theory in several ways.

As he delved deeper into the workings of MacCullagh's medium, Larmor became convinced that it could be made to do far more than even FitzGerald had realized. In particular, he found that the peculiar resistance the medium offered to the absolute rotation of its elements did not interfere with its bodily flow, which suggested that it could support permanent vortex rings of the kind William Thomson had earlier proposed might constitute material atoms.[6] While visiting Dublin early in

4. Eddington 1942: 204.

5. Larmor 1929, 1: 310–55 [1893]; though not a commissioned "report," this was given the compliment of being published among them in the annual British Association volume.

6. See Smith and Wise 1989: 417–25 on William Thomson's vortex atoms. Rotational elasticity and bodily flow are in fact independent only for very small displacements, so that a long-continued circulation, such as that around a permanent magnet, would eventually make Larmor's ether lock up. See Heaviside to FitzGerald, [Mar. 1895], FG-RDS, and FitzGerald to Larmor, 18 Mar. 1895, JL-RS. Larmor never found a fully satisfactory answer to this objection; see Larmor 1929, 2: 42–44 [1897], and Larmor to Lodge, 21 Dec. 1905, OJL-UCL.

October 1893, Larmor mentioned his discovery to FitzGerald, who immediately passed word of this "decided advance" to Lodge. "Larmor has been working at M'Cullagh and Maxwell's medium," FitzGerald reported, "and has observed that vortex rings are possible in it and this gives a delightful hook to hang a theory of the connection between matter and ether on." Larmor had "only been at it a fortnight," FitzGerald said, but had already "written a lot of his paper about it."[7]

Larmor gave more details when he wrote directly to Lodge a few weeks later. "I can correlate most things in one scheme," he said, "if I am allowed that magnetic force is velocity of the aether. Atoms are then vortex rings, and the strength of the rings is their magnetic moment."[8] Since FitzGerald had already shown that identifying magnetic force with a flow of ether and electric force with a twist made Maxwell's equations for free ether the same as MacCullagh's, it seemed to Larmor that introducing vortex atoms into a rotational ether offered a natural way to put matter and the electromagnetic field on a unified dynamical basis. Indeed, such vortex rings, conceived as vacuous cores surrounded by toroidal flows of ether, seemed to correspond beautifully to the molecular current loops Ampère had long before postulated to account for the magnetic properties of matter.

Larmor's excitement about this discovery showed itself in his first letters to Heaviside. Early in October 1893, Heaviside had written to Larmor (whom he did not then know) to ask about a recent paper of his on the electromagnetic wave surface. They exchanged several letters on the subject, but Larmor soon made it clear that his mind was elsewhere:

> I fancy I have got a grip of the aether, and I am full of the matter. I begin by establishing in its entirety MacCullagh's theory of light, then include all electrical phenomena with the usual sort of reservations or assumptions as regards dissipative actions, and end with Lord Kelvin's vortex theory of matter. Only gravitation stands now outside, and serves as a very useful *deus ex machina* to illuminate one knotty point. I wish I had nothing else to do until Christmas.[9]

Heaviside was not as enthusiastic. He had examined the rotational ether himself in 1891, and though he found that it worked well for purely field

7. FitzGerald to Lodge, 13 Oct. 1893, OJL-UCL.
8. Larmor to Lodge, 1 Nov. 1893, OJL-UCL. Larmor always used the spelling "aether," despite FitzGerald's complaints about "that inconvenient diphthong"; see FitzGerald 1902: 511 [1900], in a review of Larmor's *Aether and Matter*.
9. Larmor to Heaviside, 12 Oct. 1893, OH-IEE. Larmor thought the (unspecified) mechanism of gravitation might provide a "framework" against which the rotational elasticity of the ether could react; see Larmor 1929, 1: 409 [1893].

phenomena, he could not find any consistent way, even with additional assumptions, to make it account for electrical conduction; for the dissipative term always entered the equations in the wrong place. He cheerfully told Larmor, "Even if this rotational ether is not the thing, it is the next best thing," but he clearly thought it was a dead end.[10] Heaviside regarded Larmor as something of an interloper, a Cambridge mathematician who had wandered into Maxwellian theory without really understanding what it was about.[11] Larmor responded defensively to Heaviside's skepticism, and the cooling in the tone of the letters they exchanged over the next few weeks was the first sign of the deeper divisions that later developed between them.

Larmor sent his completed "Dynamical Theory of the Electric and Luminiferous Medium" to the Royal Society in mid-November and formally presented it at the meeting of 7 December 1893. A long abstract appeared in the *Proceedings of the Royal Society* a few weeks later and was reprinted in *Nature;* it gives a much better indication of the state of Larmor's thinking at the end of 1893 than does the heavily revised full paper that finally appeared in the *Philosophical Transactions* nearly a year later.[12] Larmor's picture of a rotationally elastic ether filled with vortex rings and lines of magnetic flow was generally well received at the meeting, though Lord Kelvin (as William Thomson had then become), who was presiding, raised some objections that later proved important.[13] Afterward Larmor had dinner with Lodge and John Perry, followed, as Perry told FitzGerald, by "interesting conversation till about 3 o/c in the morning." Perry said he only "understood about one sentence in ten" of this conversation, but both he and Lodge were impressed by Larmor's ideas. Lodge declared himself "fascinated" and "much excited" by the theory; it was, he told Larmor, "a capital work for the century to die with."[14]

For all its fascination, however, Larmor's theory already faced serious problems, as Lodge was well aware. Larmor had in fact first written to Lodge (whom he then knew only slightly) late in October 1893 to ask

10. Heaviside to Larmor, 6 Dec. 1893, JL-RS; cf. Heaviside 1893–1912, 1: 127–31 [1891].

11. See Heaviside to FitzGerald, 4 Feb. 1895, FG-RDS.

12. Cf. the abstract, Larmor 1929, 1: 389–413 [1893], with the full paper, 1: 414–535 [1894]; the abstract also appeared in *Nature 49* (1894): 260–62, 280–83. In what follows, I draw freely on the more detailed account of the evolution of Larmor's theory in Buchwald 1985a: 133–73.

13. See Larmor to FitzGerald, 8 Dec. [1893], FG-RDS, a fragment that ends: "Kelvin was my only critic: he put several points which puzzled. . . ."

14. Perry to FitzGerald, 16 Dec. 1893, FG-RDS; Lodge to FitzGerald, 19 Dec. 1893, FG-RDS, and Lodge to Larmor, 8 and 11 Dec. 1893, JL-RS.

about a possible experimental test of his theory. Lodge had sent Larmor a copy of his paper "Aberration Problems" a short time before, and Larmor wondered whether Lodge's interferometer apparatus could be used to look for the acceleration of light along lines of magnetic force (and thus ether flow) that his theory predicted. Lodge replied that he had already looked for such an effect and found nothing but would now try a more careful test. After running beams of light along the axes of a set of magnetized coils, he was soon able to report that the magnetic flow, if it existed at all, must be very slow.[15] Larmor was clearly disappointed; he told Heaviside, "My chief trouble is that Lodge cannot find any effect of a magnetic field on the velocity of light" and confessed to FitzGerald that it was "the weak(est) spot I think in the whole affair."[16] He reported Lodge's negative result in a brief postscript to his Royal Society paper but tried to explain it away by suggesting that the ether might simply be extremely dense—at least as dense as ordinary matter—and its rate of flow correspondingly slow. Larmor himself called this a "somewhat startling" hypothesis, but he was willing to accept it as the only apparent way to save his theory from experimental refutation.[17]

Larmor also had to deal with the objection Kelvin had raised at the meeting: the force between his vortex rings was in the wrong direction. For closed vortex rings to correspond to Amperean current loops as Larmor had suggested, rings whose flow was in the same direction would have to repel each other, just as similar magnetic poles do. But Kelvin had shown long before that closed rings of the same sign in fact *attracted* each other, simply in consequence of the laws of vortex mechanics, a point Larmor had somehow overlooked. "Kelvin's paradox," as Jed Buchwald has called it, was rather subtle, and it took Larmor some months to grasp its full implications.[18] When he finally realized how deeply it cut into his attempt to apply the rotational ether to material media, he was forced to change his theory in fundamental ways, as we shall see.

Larmor's paper reached its first Royal Society referee, J. J. Thomson,

15. Larmor to Lodge, 28 Oct. 1893, OJL-UCL, and Lodge to Larmor, 3 Nov. 1893, JL-RS; cf. Lodge to Larmor, 29 Oct., 1 and 2 Nov. 1893, JL-RS, and Lodge notebook 3.17, pp. 100–104 (dated 3 Nov. 1893), OJL-UL, describing the experiment and concluding that the flow, if any, must be slower than 30 cm./sec. See Hunt 1986: 124–32 for a more detailed account of this episode.

16. Larmor to Heaviside, 22 Dec. 1893, OH-IEE; Larmor to FitzGerald, 18 Dec. 1893, JL-RS.

17. Larmor 1929, 1: 413 [1893] and 1: 483 [1894].

18. See Buchwald 1985a: 154–61 and Smith and Wise 1989: 424. After voicing his objection at the 7 Dec. 1893 meeting, Kelvin explained it more fully in a letter to Larmor, 26 Dec. 1893, JL-RS; cf. Larmor 1929, 1: 505–6 [1894].

in mid-December. Thomson was daunted by the length and ambitiousness of the paper—it constituted, he said, "a kind of Physical Theory of the Universe"—and when he finally submitted his report in February 1894, he confessed that he had found parts of Larmor's argument hard to follow, a remark later readers have echoed. Though he agreed that the rotational ether gave a good account of free fields, Thomson did not think Larmor's attempt to extend it application to material media had succeeded; the paper did not, he said, "give any help in gaining a clear insight into what goes on in any physical process in which matter as well as ether is concerned." Nonetheless, he recommended that the paper be published in full and suggested no major revisions.[19]

Lodge had initially been asked to serve as the second referee but suggested that the paper be sent to FitzGerald instead, "lest," as he said, "I didn't understand it all."[20] There could have been no better choice, not only because of FitzGerald's unrivaled grasp of both MacCullagh's and Maxwell's theories but because of his willingness to act more as a collaborator than a referee. Shortly after the paper reached him early in February 1894, he requested and received permission from the Royal Society to break the usual rule of anonymity and discuss it directly with Larmor.[21] He proceeded to subject the paper to detailed and searching criticism, first in a series of long letters between late February and mid-April and then in another intense exchange in July. Working behind the scenes, FitzGerald played a crucial role in drawing out the implications of Larmor's theory and in steering it toward its final form. Heaviside was not far from the mark when, after reading the completed paper, he told FitzGerald he was "inclined to think you are the virtual author of a good bit in L[armor]'s memoir."[22]

In fifteen years, the Maxwellian story had come almost full circle. In February 1879, Maxwell had completed his referee report on FitzGerald's electromagnetic interpretation of MacCullagh's theory; now, in February 1894, FitzGerald was beginning his own examination of Larmor's paper on virtually the same topic. Maxwell's theory had gone through important changes in this period, to most of which Larmor seemed quite oblivious. His work was rooted in an older Cambridge

19. J. J. Thomson, [report on Larmor, "Dynamical Theory"], 5 Feb. 1894, RR.12.160, RS Archives; cf. Buchwald 1985a: 162.

20. Lodge to FitzGerald, 19 Dec. 1893, FG-RDS.

21. FitzGerald to Rayleigh, 9 Feb. 1894, RR.12.161, RS Archives; Rayleigh to FitzGerald, 11 Feb. 1894, FG-RDS.

22. Heaviside to FitzGerald, 4 Feb. 1895, FG-RDS; cf. Heaviside to FitzGerald, 28 Jan. 1895, FG-RDS, asking for FitzGerald's assessment of Larmor's paper, and FitzGerald to Heaviside, 1 Feb. 1895, OH-IEE, explaining the revisions he had urged on Larmor.

tradition of generalized dynamics and elastic solid theories and had little to do with the emphasis on energy flow and the simplified formalism that had dominated Maxwellian theory since the mid-1880s. But the tradition Larmor represented was not yet entirely played out, and when combined with the newer Maxwellianism, in the person of FitzGerald, it was still able to produce some very fruitful results.

Inventing Electrons

FitzGerald found much to criticize in the initial version of Larmor's theory, particularly in its treatment of conduction and other phenomena involving material media. Larmor seemed to have little feel for the realities of force and motion that underlay the electromagnetic equations; he often glossed over important details and seemed unable to distinguish plausible mechanisms from implausible ones.[23] FitzGerald undertook to supply this missing physical insight, and over the spring and summer of 1894 he pushed Larmor into making a series of revisions aimed at clarifying and correcting the physical content of his theory.

Larmor refused to yield to FitzGerald on one important point: the structure of the rotational ether itself. FitzGerald argued that its resistance to absolute twist implied an underlying mechanical structure, either an array of gimbaled gyrostats like the one William Thomson had devised in 1889 or a tangle of filaments like the vortex sponge Fitz-Gerald had himself proposed in 1885. Larmor insisted, however, that rotational elasticity might simply be one of the ultimate properties of the ether and that the potential functions and the principle of least action provided a sufficient dynamical foundation for his theory. FitzGerald eventually conceded the point, albeit reluctantly; as he later told Heaviside, the question of ultimate properties was "semimetaphysical" and more a matter of philosophy and scientific taste than of ascertainable fact.[24] But FitzGerald's own tastes ran strongly toward mechanical explanations, particularly ones framed in terms of the vortex sponge, and he never regarded Larmor's rather abstract dynamical ether as an altogether satisfactory foundation for his theory.

FitzGerald also raised questions about the way Larmor handled electrostatic forces. Having abandoned the usual "Maxwell stress" in the ether—a tension along the lines of force combined with a lateral pres-

23. Heaviside thought Larmor lacked "any natural sense of dynamics"; see Ludwig Silberstein to Heaviside, 10 Jan. 1917, OH-IEE, where this remark is quoted back to him.
24. FitzGerald to Heaviside, 1 Feb. 1895, OH-IEE; cf. Larmor to FitzGerald, 4 Apr. 1894, FG-RDS, and Larmor 1929, 1: 519–21 [1894].

sure—in favor of a purely rotational stress, Larmor had to find some other way to derive the known forces between charged bodies. He responded with an elaborate and rather fanciful mechanism in which the motion of charged matter through the ether somehow generated electromagnetic "wavelets" that built up to yield appreciable forces on other charged bodies. This "explanation" satisfied no one, however, not least because it yielded no force at all on stationary charges; Heaviside dismissed it as *"rather thin,"* while FitzGerald said he could not "make head or tail of it."[25] Larmor's treatment of ordinary conduction currents was also less than compelling, largely because he had no real answer to Heaviside's demonstration that true conductivity was impossible in a rotational ether. Left with no middle ground between the absolute nonconductivity of his ether and the perfect conductivity of his vortex rings, Larmor was reduced to invoking a mysterious intermittent "rupture" of the ether between the rings as the only way to account for the finite conductivity of ordinary circuits and the consequent dissipation of energy within them.[26]

As FitzGerald raised more and more objections to his theory, Larmor responded by piling on more and more complications, especially in his successive attempts to dispose of "Kelvin's paradox." In the closed vortex rings that he had originally equated with current loops, the field had no way to get a "grip" on the core of the vortex; this made it impossible for the field to generate or alter currents by electromagnetic induction and also resulted in mechanical forces of the wrong sign. Larmor managed to evade the problem for macroscopic currents by postulating that they always flow in open circuits, completed by displacement, convection, or discharge processes across nonconducting gaps like those between the plates of a condenser, the poles of a battery, or the molecules of an ordinary conductor.[27] The rotational elasticity within such gaps enabled the field to grip and alter the current there, thus making induction possible and yielding forces of the proper sign. But Larmor was vague about what went on in these nonconducting gaps, particularly the "rupture" of the ether that was necessary for the actual flow of current. Worse, he was still left with forces of the wrong sign between permanent magnets, since he had found no way to "open" his Amperean loops without destroying the permanence of their currents.

25. Larmor 1929, 1: 402 [1893]; cf. Heaviside to FitzGerald, 13 Apr. 1894, FG-RDS, and FitzGerald to Heaviside, 1 Feb. 1895, OH-IEE.

26. Larmor 1929, 1: 406 [1893]; cf. Larmor to Heaviside, 10 Jan. 1894, OH-IEE, and Heaviside to FitzGerald, 13 Apr. 1894, FG-RDS. See also Buchwald 1985a: 146–49.

27. Larmor 1929, 1: 400–401 [1893]; Larmor introduced these open circuits in a note added to his abstract on 7 Dec. 1893, which suggests he had not at first appreciated that closed vortex rings would not behave like closed electric currents.

Though his objections had by no means all been answered, FitzGerald approved Larmor's paper for the Royal Society in April 1894, contingent on the inclusion of some revisions and additions.[28] Larmor got the paper back in mid-May and spent the next few weeks "polishing occasionally in my spare time," as he told FitzGerald. He was "still in the Serbonian bog as regards ponderomotive forces," he said, particularly the old puzzle of the force between closed Amperean rings: "I cannot make up my mind for certain how far I must give up free magnetism."[29] He said he hoped to discuss this further when FitzGerald came to Cambridge as an examiner early in June, and it was soon after this visit that he finally sent the Royal Society his revisions and a long addendum to his paper.

Large parts of Larmor's original optimistic plan for a unified theory of ether and matter had by then collapsed, and the June addendum was largely an effort to salvage something from the wreckage. Most of it was devoted to yet another attempt to escape from Kelvin's paradox, this time by an elaborate mechanism involving the thermal motion of the vortex rings. This proved no more satisfactory than his earlier efforts, however, and in the end Larmor had to admit that "in unravelling the detailed relations of aether to matter," his theory was "not very successful."[30]

Ironically, by the time he sent in this addendum, Larmor had already hit on the idea that was to reverse the fortunes of his theory and give it its lasting value. He had realized from the first that his rings could support charge as well as current; that is, they could be surrounded not only by the vortical flow of ether that corresponded to the magnetic field around a current but also by the pattern of rotational strain that corresponded to the electric field around a static charge. But Larmor had given these "atomic charges" only a minor role in his theory; their electrification, which he treated as spread evenly over the entire core of the vortex, served as little more than a convenient "glue" with which to hold several rings together to form a complex molecule.[31] In late March and early April, however, FitzGerald had kept prodding Larmor to put something *inside* his vortices for the field to grab onto.[32] Only in this way, he had argued, could the Amperean current loops be opened up and Kelvin's paradox finally be overcome. As Larmor explored this idea over the next few weeks, he found that the charge he had formerly spread

28. FitzGerald to Rayleigh, 19 Apr. 1894, RR.12.162, RS Archives.
29. Larmor to FitzGerald, 25 May 1894, FG-RDS.
30. Larmor 1929, 1: 512 [1894]; Larmor's "baroque mechanism" is discussed in Buchwald 1985a: 150–53.
31. See Larmor 1929, 1: 401 [1893] and 1: 468, 474 [1894].
32. FitzGerald to Larmor, 30 Mar. and 3 Apr. 1894, JL-RS.

over the surface of his rings could be broken up into pointlike foci of rotational strain that provided just the sort of anchors for the field that FitzGerald had called for. Moreover, these charged "monads," as he initially called them, seemed to offer solutions to many of the other problems that had plagued his theory. Pushed along by FitzGerald, Larmor had, in effect, invented the electron.

Larmor first broached his new idea in letters to Lodge and FitzGerald at the end of April and in May, but he barely mentioned it in the revisions he sent the Royal Society in June.[33] Only after considerable exploratory work and another intense bout of correspondence with FitzGerald in July did he feel ready to commit himself in print to a theory based on what he now called "electrons." (The word had been coined in 1891 by G. J. Stoney and came to Larmor via FitzGerald, who wrote on 19 July 1894 to say that Stoney had been visiting him and "was rather horrified at calling these ionic charges 'ions.' He or somebody has called them 'electrons' and the ion is the atom not the electric charge.")[34]

In August, Larmor sent the Royal Society a second long addendum, headed "Introduction of Free Electrons," in which he summarized the discoveries he had made over the last few months and sketched out much of his later work.[35] At a stroke, he rendered obsolete the theory of vortex rings he had so laboriously constructed since the previous October and showed that all he had tried to do with the rings could be accomplished more easily and naturally by freely moving electrons. Electrons offered an answer not only to Kelvin's paradox but to Heaviside's more sweeping objection to any true conductivity in a rotational ether: they allowed Larmor to dispose of "such a barbarous makeshift as rupture of the aether," as he now called it, and to treat all conduction currents as essentially convective.[36] Conductivity in the Maxwellian sense simply dropped out of his theory, as he now traced the flow of ordinary currents and the dissipation of their energy entirely to the motions and collisions of electrons. Similarly, he began to treat displacement in material dielectrics as a real polarization of molecules by the separation of the positive and negative electrons within them. Larmor's

33. Larmor to Lodge, 30 Apr. 1894, OJL-UCL; Larmor to FitzGerald, 25 May 1894, FG-RDS; Larmor 1929, 1: 468n, 474–75 [both added 14 June 1894].
34. FitzGerald to Larmor, 19 July 1894, JL-RS. On Stoney and the electron, see O'Hara 1975. Larmor began to use the word immediately (see Larmor to Lodge, 30 July 1894, OJL-UCL), and other theorists took it up from him in the late 1890s. At first it was applied to both positively and negatively charged particles, but its use was gradually restricted to J. J. Thomson's negatively charged "corpuscles."
35. Larmor 1929, 1: 514–35 [added 13 Aug. 1894].
36. Larmor to Lodge, 30 Apr. 1894, OJL-UCL.

characteristic strategy, and his great departure from traditional Maxwellian theory, was to treat the macroscopically defined electric and magnetic properties of matter (its conductivity, permittivity, and permeability) as *consequences* of its electronic microstructure. Indeed, he proposed to build up matter itself wholly from orbiting swarms of electrons and so to reduce the entire physical universe to simply a collection of singularities in the ether.

Larmor knew how bold an idea this was, and he ended a letter to FitzGerald by saying, "I must speculate no more for fear people begin to speculate whether I am sane."[37] But as he pushed his speculations further, he found that the introduction of the electron cleared up many persistent puzzles in electrodynamics and optics in a neat and seemingly natural way. The optimism with which he had begun in October 1893, and which had sunk so low by the following spring, returned and rose even higher in the latter half of 1894. By March 1895 his enthusiasm was almost unbounded; "unless I have blundered," he told Lodge, he had found "an actual dream of a system that goes like the world."[38]

When Larmor's "Dynamical Theory" finally appeared in the *Philosophical Transactions* at the end of 1894, it was layered over with corrections and additions and already largely out of date. It was followed in 1895 by "Part II," in which he clarified the foundations of electron theory and applied it to optical phenomena; in 1897 by "Part III," in which he treated material media in more detail; and finally in 1900 by *Aether and Matter*, in which he gave a comprehensive account of his theory of electrons, with little reference to the rotational ether with which he had begun.[39] Larmor's papers, and especially his book, had a profound effect on the rising generation of British theorists. Years later, Arthur Eddington told Larmor that *Aether and Matter* had "opened [his] eyes to the new line of thought" when he was a student; "I and my contemporaries spent many hours trying to explain to each other what on earth it meant," he said, and they had absorbed electron theory largely from Larmor's writings.[40]

Larmor's later work was strongly influenced by Lorentz's parallel researches, particularly his *Versuch einer Theorie der electrischen und optischen Erscheinungen in bewegten Körpern*, which Larmor read shortly after its

37. Larmor to FitzGerald, 7 July 1894, FG-RDS.
38. Larmor to Lodge, 29 Mar. 1895, OJL-UCL.
39. Larmor 1929, 1: 543–97 [1895] and 2: 11–132 [1897]; Larmor 1900. *Aether and Matter* was a revised version of Larmor's 1898 Adams Prize essay.
40. Eddington to Larmor, 14 Oct. 1932, JL-RS. See Warwick 1989 on the later work of Larmor and his students.

publication early in 1895.[41] Like most Continental physicists, Lorentz had come to Maxwell's theory through Helmholtz's version of it, and he did not grasp, or did not accept, the Maxwellian view of charge as simply a manifestation of discontinuities in the displacement in the field. Instead, he merged Maxwell's theory of interaction via fields with the Weber-Clausius theory of particulate charges, and in 1892 he produced the first version of what was later to become his electron theory.[42] Unlike Larmor, Lorentz did not try to explain electrons as structures in the ether but simply posited the existence of tiny charged particles within material bodies. The sharp distinction he drew between ether and matter (that is, between fields and electrons) enabled Lorentz to construct a clear and remarkably effective theory of electrodynamics, optics, and the structure of matter and to treat phenomena of moving media which had defeated the old Maxwellian single-medium approach.

As the work of Lorentz and Larmor came together around 1895, electron theory emerged as one of the most active and promising areas in all of physics—yet it remained largely hypothetical. The experimental evidence for the actual existence of electrons was still indirect and based mainly on refined analyses of subtle optical effects. But from the mid-1890s, more direct evidence that matter really did contain swarms of tiny charged particles began to accumulate. Larmor had suggested as early as 1894 that cathode rays might consist of streams of his "monads," and this speculation was borne out over the next few years by the experiments of J. J. Thomson and others.[43] The idea that atoms contained orbiting electrons also received striking support late in 1896 when Lorentz's former student Pieter Zeeman observed the splitting of spectral lines in a magnetic field. Zeeman's discovery was soon confirmed and extended by Lodge and by FitzGerald's protege Thomas Preston, and the analysis of the effect by Lorentz and Larmor not only showed that it was due to orbiting negative electrons but yielded a ratio of charge to mass very close to that later found in cathode ray experiments.[44] By 1897, or 1900 at the latest, most physicists accepted electrons as real constituents of matter, and a growing number were willing to make them the foundation of a new view of the physical universe.

41. Lorentz 1895. See Larmor to Lodge, 29 Mar. and 18 Apr. 1895, OJL-UCL, on Larmor's reading of Lorentz.

42. On Lorentz's electron theory, see Jungnickel and McCormmach 1986, 2: 231–36, and the sources cited there.

43. Larmor to FitzGerald, 15 July 1894, FG-RDS; Larmor to Lodge, 30 July [1894], OJL-UCL. Although they were both in Cambridge and worked on closely related problems, Larmor and J. J. Thomson had little contact in the 1890s.

44. Whittaker 1951: 411–16; Weaire and O'Connor 1987.

"Larmor's Force"

The success of electron theory raised the question of how it fit with existing Maxwellian theory. Could electrons simply by grafted onto the existing theory, or would substantial parts of the old growth first have to be cut away? This remained an open question through the latter half of the 1890s, and the debates surrounding it served to highlight important differences in the ways both Maxwellian theory and the new electron theory were conceived.

The issue first became acute early in 1895, when Larmor claimed that he could prove that Maxwell's theory of currents was wrong and that currents *must* consist of streams of discrete electrons. He had reached this conclusion by a rather roundabout route. A key step in Maxwell's derivation of the dynamical relations between currents in his *Treatise* had been his analysis of circuits into discrete elements, each of which he treated as controlling a definite portion of energy in the surrounding field.[45] In his August 1894 addendum on "Free Electrons," Larmor had taken moving electrons as equivalent to elements of current and, transposing these into Maxwell's circuit elements, had used the vector potential and the principle of least action to derive the force between two currents. But instead of the purely transverse force given by Ampère and Maxwell, Larmor found the total force on an element of current **u,** taken as lying along the x axis, to be $(udF/dx, udF/dy, udF/dz)$, where F is the x component of the vector potential **A** (G and H being its y and z components). This force differed from the Ampère force on such an element, $\mathbf{u} \times \mathbf{B}$ (equal to $\mathbf{u} \times$ curl **A**), by an amount $(udF/dx, udG/dx, udH/dx)$, soon dubbed "Larmor's force."[46] This additional force, the product of the current and the slope of the vector potential along its direction, summed to zero around a closed circuit, which is perhaps why Maxwell had ignored it; nonetheless, it would produce readily detectable effects in a nonrigid conductor such as a filament of mercury. Indeed, Larmor said that it should "admit of easy test."[47]

When Larmor's paper finally appeared at the end of 1894, his prediction of a new force attracted the attention of experimentalists, and Perry, Lodge, and FitzGerald's colleague John Joly soon set about devising ways

45. Maxwell 1873, art. 586; cf. Buchwald 1985a: 54–59, 168–71.
46. Larmor 1929, 1: 527–29 [1894]; FitzGerald to Lodge, 21 Feb. 1895, OJL-UCL. FitzGerald remarked to Heaviside, 8 Mar. 1895, OH-IEE, that Larmor's force "must be wrong for its vector expression is bothersome, I can't see any good way of expressing it without resolving it."
47. Larmor 1929, 1: 529 [1894].

to detect it.[48] A flurry of correspondence ensued in February and March 1895 as Lodge and FitzGerald pressed Larmor to explain his ideas more fully. FitzGerald soon began to express doubts about both the reality of the new force and the validity of Larmor's reasoning. He was especially critical of the way Larmor had used the vector potential, telling him repeatedly that since it "does not *locate* the energy properly" in the field, it was not likely to yield the correct force on an element of current. The product of the current and the vector potential "may *not* be a true representation of the kinetic energy of an *element* of current," FitzGerald said, "and only, as it were accidentally, gives right results because all its mistakes cancel out when taken round a closed circuit."[49] To get at the forces on *parts* of a circuit, one should start with the true distribution of energy in the surrounding field, given by $\frac{1}{2}\mu H^2$; only by focusing on the state of the field at each point, FitzGerald insisted, could one hope to find the forces that would be exerted there.

At the end of February, FitzGerald performed a simple experiment with magnets and a mercury trough which effectively proved that Larmor's force did not exist; Lodge followed a few weeks later with another showing the same result.[50] But it was a thought experiment FitzGerald described to Larmor on 7 March that brought out the flaws in Larmor's reasoning most forcefully and, FitzGerald said, "ultimately shook him."[51] Picture a ring of wire centered on the x axis and carrying a steady current (see Figure 9.1). Its magnetic field is shown by dotted lines. The lines of vector potential form circles parallel to the ring, and so have no x component; F is zero everywhere, as are dF/dx, dF/dy, and dF/dz. Now picture a steady current running parallel to the x axis, so that its only component is $\mathbf{u}$. Each of the terms of Larmor's expression for the total force is thus zero, indicating no net force on any segment of the current; Larmor's new force just cancels the ordinary Ampère force. But it was

48. On Perry, see Larmor to FitzGerald, 16 Feb. 1895, FG-RDS; on Joly and Lodge, see FitzGerald to Larmor, 18 Feb. 1895 (in sheet 470.3, misfiled as part of a letter of 11 Feb.) and 21 Feb. 1895, JL-RS.

49. FitzGerald to Larmor, 21 and 23 Feb. 1895, JL-RS. FitzGerald made similar remarks in his letters of 18 Feb., 7, 10, and 15 Mar. The claim in Buchwald 1985a: 58–59 that "Maxwell and most Maxwellians" (except Heaviside) regarded the vector potential form of the energy density as "the fundamental expression" does not apply to FitzGerald, at least by this time, and its application even to Maxwell seems questionable; see Maxwell 1873, art. 636.

50. FitzGerald to Larmor, 2 Mar. 1895, JL-RS; Lodge to Larmor, 2 Apr. and 5 May 1895, JL-RS.

51. FitzGerald to Heaviside, 15 Mar. 1895, OH-IEE. FitzGerald also showed that Larmor's force would lead to violations of the conservation of energy; see FitzGerald to Larmor, 27 Feb. and 10 Mar. 1895, JL-RS.

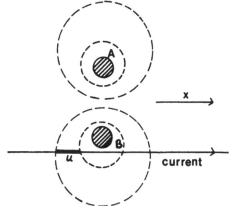

9.1. FitzGerald's March 1895 diagram of a thread of mercury carrying a current in a magnetic field, showing the nonexistence of "Larmor's force." The magnetic field around a ring of current-carrying wire (shown in cross-section as *A* and *B*) will blow the thread of mercury *u* out of the plane of the paper, contrary to Larmor's prediction that a compensating force would prevent any motion.

perfectly obvious, FitzGerald said, that the magnetic field *would* exert a force on his segment of current and that "if it were a bit of a blob of mercury on a glass plate it would be blown off the plate"—so obvious that the experiment was hardly worth doing. Larmor's force and the reasoning that had led him to it were, FitzGerald declared, "almost certainly wrong."[52]

Although he admitted that FitzGerald's letter "at first glance took my breath away," Larmor answered that he was "not blown out of the water, along with that blob of mercury of yours, just yet."[53] He was convinced that his force followed necessarily from any theory based on current elements and thus a fortiori from his electron theory; giving up his force would, he thought, mean giving up his electrons as well—something he was by then very reluctant to do. But FitzGerald's arguments were overpowering, and by the second week of March it was clear that something had to give. On reconsidering the whole problem, Larmor concluded that he had been wrong to identify his moving electrons with current elements as traditionally conceived. His earlier analysis would in fact apply only if the electrons were fixed to the wire and dragged together through the ether, he said, and he now found that electrons left free to

52. FitzGerald to Larmor, 7 Mar. 1895, JL-RS. Sheet 465.2 of this letter has been misfiled as part of a letter of 18 Jan. 1895; it is quoted with this mistaken date in Buchwald 1985a: 170, resulting in a jumbled chronology.
53. Larmor to FitzGerald, 9 Mar. 1895; the first (quoted) part of this letter is in JL-RS and the rest in FG-RDS. Larmor tried to escape from FitzGerald's argument by invoking a couple which, by his analysis, would also act on each element of the current, but FitzGerald pointed out that this would merely set the mercury spinning, not blow it away; see FitzGerald to Larmor, 10 Mar. 1895, JL-RS.

move on their own would be subject to no additional forces, just the Ampère force "pure and simple."[54] His electron theory was saved.

But Larmor did not stop there: he still maintained that Maxwell's own methods, if properly applied, necessarily led to the discredited tension. Maxwell's theory of currents had been accepted for so long, he said, only because no one had rigorously examined its implications. Larmor now reversed the apparent verdict of the experiments: far from proving the electron theory of currents to be false, the observed absence of any tension in fact proved it to be the only one possible. When he wrote up his account of the episode for Part II of his "Dynamical Theory" in May 1895, he presented it as a triumph for his electron theory and a defeat for the old Maxwellian approach.[55]

The Maxwellians, notably Heaviside and FitzGerald, did not see it quite that way. Larmor had followed the original treatment in Maxwell's *Treatise* quite closely—too closely, they thought. His conclusions depended crucially on the use of current elements, the vector potential, and the principle of least action—all of which Heaviside and FitzGerald had by then abandoned except as mathematical artifices. "I'm afraid that poor dear convenient vector potential has much to answer for," FitzGerald said, and both he and Heaviside regarded it as the source of most of Larmor's errors.[56] Heaviside told FitzGerald that Larmor's use of "the vector potential in an expression for the *mech[anical] force* (something real) is a sufficient condemnation, I think. The mech[anical] force must depend solely on the state of things at the place; the v[ector] p[otential] *does not,* so action at a distance is involved in his assumption unless he is prepared to elaborate a theory in which A is physically significant as a state of the medium; not merely a coordinate."[57] FitzGerald agreed and told Larmor that, in using the vector potential and related formulas, he was "going back on the old vomit of action at a distance for the sake of analysis."[58]

As for current elements, FitzGerald and Heaviside had discarded

54. Larmor to FitzGerald, 12 Mar. 1895, FG-RDS; cf. Larmor 1929, 1: 537–38 [1895]. Buchwald 1985a: 171 says that in abandoning current elements, "Larmor had finally abandoned Maxwellian theory"; I would instead say (following my own usage of "Maxwellian") that Larmor had finally caught up with the leading Maxwellians in abandoning an inappropriate part of Maxwell's original formulation.

55. Larmor to FitzGerald, 18 Mar. 1895, FG-RDS; Larmor 1929, 1: 547–48 [1895]. In the expression near the bottom of p. 547, the coefficients of the first two components of the force should be w, not u and v; cf. Larmor 1900: 338.

56. FitzGerald to Heaviside, 8 Mar. 1895, OH-IEE.

57. Heaviside to FitzGerald, [Mar. 1895], FG-RDS. Internal evidence places this undated fragment between FitzGerald's letters to Heaviside of 8 and 15 Mar. 1895, OH-IEE.

58. FitzGerald to Larmor, 15 Mar. 1895, JL-RS; cf. [FitzGerald] 1898a, a proof copy of which is in FG-RDS.

them years before. It was a mistake, FitzGerald told Larmor, to treat "elements of currents as if they were complete entities while we know that they are only the cores of a complex phenomenon"; indeed, the whole idea of current elements did not fit with the basic Maxwellian view of currents as little more than reflections of activity in the surrounding field.[59] "The whole method is fundamentally wrong," Heaviside had said in 1888, "and of little practical service in the investigation of electromagnetism from the physical side, i.e., with propagation in time through a medium." Current elements, he declared, simply "are *not in it*."[60]

Even the principle of least action, the foundation of Larmor's whole approach, did not escape unscathed. Heaviside had long regarded it more as a source of confusion and mystification than as the powerful tool of discovery its adherents claimed it to be. It was all very well to deduce the behavior of a dynamical system from expressions for the kinetic and potential energies, T and U—but how was one to find those expressions in the first place? As Heaviside remarked to FitzGerald, "You can only come to the notion that T is kin[etic] energy by a very considerable study of the facts of the case, which involves a knowledge of the forces, mechanical and electromotive"; and once those forces were found, it seemed simpler to use them directly than to embark on a long detour through Lagrangian functions and generalized coordinates.[61] The principle of least action had been made into "a golden or brazen idol," Heaviside later said, and though "the mathematical tutors and lecturers at Cambridge . . . make the young men fall down and worship the idol," it was of little use in solving real electrical problems. "I have never practised it myself," Heaviside said, "(except with pots and pans)."[62] FitzGerald came to share Heaviside's view, calling the principle of least action an "analytical juggle" and arguing that it could not provide a proper foundation for physical theory.[63]

Larmor ignored most of the changes Maxwell's theory had undergone since 1873; instead of focusing on the localization and flow of energy in the field and using the new set of "Maxwell's equations" that embodied this approach, he continued to use the more elaborate mathematical

59. FitzGerald to Larmor, 18 Mar. 1895, JL-RS.
60. Heaviside 1892, 2: 501 [1888].
61. Heaviside to FitzGerald, 17 Mar. 1895, FG-RDS.
62. Heaviside 1893–1912, 3: 175 [1903]; cf. Heaviside to Lodge, 30 Oct. 1899, OJL-UCL, where Heaviside says that the principle of least action "is a most clumsy machine in electromagnetics, but is splendid in the house."
63. FitzGerald 1902: 509 [1900], in a review of Larmor's *Aether and Matter;* cf. p. 331 [1895], where FitzGerald objected that by setting nature to minimize an integral over time, the principal of least action "makes the present depend on the future."

tools and methods long sanctified at Cambridge. Even after FitzGerald had shown him how clumsy those tools could be and how far astray they had led him over his "new force," Larmor remained reluctant to give them up; he would go no further in April 1895 than to suggest that "Maxwell's use of (F,G,H) is obscure and (perhaps) wrong in principle when displacement currents are important."[64] It was a conclusion Heaviside had reached in a much stronger form by 1885 and FitzGerald not long after. If Larmor could now prove that Maxwell's original formulation of his theory was mistaken, this was, by 1895, hardly news to the leading Maxwellians.

When Heaviside told FitzGerald early in March 1895 that "Maxwell was only $\frac{1}{2}$ a Maxwellian," it was to make the point that "L[armor] is only $\frac{1}{4}$ or less."[65] In the course of extending and revising Maxwell's theory in the 1880s, the Maxwellians had opened a gap between Maxwell's work and their own. Larmor made another new departure in the mid-1890s, but he did so by starting directly from Maxwell's own work and from older electrodynamic theories, without passing through a truly Maxwellian stage. His methods and approach were quite different from those of FitzGerald and Heaviside and their immediate followers. In assimilating electron theory, the Maxwellians drew on Larmor's achievements, but they did not follow his path.

Assimilating Electrons

For his own part, FitzGerald thought the best way to fit electrons into Maxwellian theory would be to explain both in terms of an underlying vortex ether. He had urged this approach on Larmor from the beginning, and in an 1896 lecture to the Chemical Society he went so far as to declare that "a theory of electromagnetic actions depending entirely on the actions of electrons" had already been "worked into" the vortex sponge.[66] FitzGerald had long speculated that lines of electric force were simply long vortex filaments twisted into corkscrewlike spirals, and in a paper at the Royal Dublin Society in 1898 he pointed out that a permanent kink in such a spiral, with a change in the "handedness" of the twist on either side, would correspond to a discrete charge—an electron.[67] He had no illusions that this paper was conclusive; indeed, when

64. Larmor to FitzGerald, 1 Apr. 1895, FG-RDS.
65. Heaviside to FitzGerald, [Mar. 1895], FG-RDS (see n. 57 above).
66. FitzGerald 1902: 353 [1896].
67. Ibid., pp. 472–77 [1898]; this was the first paper FitzGerald had given at the RDS since he split with it in 1889. See also FitzGerald to Heaviside, 25 Aug. 1893, OH-IEE, on vortex spirals, and FitzGerald 1902: 484–86 [1899], comparing the energy density of a vortex sponge with that of the electromagnetic field.

he sent a copy to Heaviside, he called it a "wild hypothesis in a blue wrapper" (RDS offprints had blue covers) and said, "You see what I am coming to in my old age—vague guesses that pretend to be founded on something that may deceive some people into the idea that it is mathematical reasoning."[68] But these "vague guesses" were founded on a deep faith in the vortex sponge as the ultimate basis of both ether and matter.

FitzGerald voiced this faith very clearly in an address to the Dublin section of the Institution of Electrical Engineers early in 1900. After sketching some of the functions of the ether, he said:

> We are still looking for a theory of its structure which will give a dynamical explanation of its properties. The direction in which it is most probable that an explanation will be found is in the hypothesis that the ether is of the nature of a perfect liquid full of the most energetic motion. . . . So far as this hypothesis has been worked at there seems nothing impossible about it, but, on the contrary, much possibility in it, and, to my mind, its inherent simplicity confers on it a great probability.[69]

Lodge agreed and declared, in 1907, that the ether was almost certainly a perfect liquid "squirming internally with the velocity of light."[70] Like most British physicists of their generation, FitzGerald and Lodge regarded the advent of the electron not as a threat to their view of the ether but as an extension of it. They had always looked on the laws of electromagnetism as consequences of the structure and motion of the ether, and they believed that the discovery of the electron had simply revealed another level of that structure while leaving its foundations in the ether intact.

Heaviside shared this basic premise, but in practice he paid less attention to the mechanics of the ether than did FitzGerald and Lodge. Instead of trying to build up electrons from hypothetical vortices in the ether, he simply extended the macroscopic laws of electromagnetism down to the microscopic scale. The field equations always came first for Heaviside; electrons were simply an additional hypothesis to be accommodated by adjusting the boundary conditions. His pioneering work on the fields around moving charges was all based on this approach, and he saw no reason why the advent of the electron should necessitate any fundamental changes in electromagnetic theory.

Heaviside first encountered electron theory in 1893, well before Larmor hit on the idea. In 1891, Lodge had written to Lorentz to ask about some of his recent work on aberration. Lorentz replied with a three-

68. FitzGerald to Heaviside, 7 and 20 May 1899, OH-IEE; cf. FitzGerald to Kelvin, 9 Jan. 1899, WT-ULC.
69. FitzGerald 1902: 490 [1900].
70. Lodge 1909: 97 [1907]; cf. Lodge 1931a: 85.

page handwritten abstract in English of his first important paper on electron theory, which was not published in full (in French) until the next year. In November 1893, a question about Fresnel's drag hypothesis prompted Lodge to send this abstract to Heaviside, who returned it a few days later with his comments.[71] After translating Lorentz's equations into his own notation, Heaviside told Lodge the theory contained little that was really new; it was simply "Maxwell's theory limited to a particular state of things," with "a lot of electrification moving about" in the ether, but with no true conductivity. He saw parallels to his own work on moving charges, but he thought Lorentz had been both too hypothetical in positing the actual existence of charged particles in matter and too restrictive in excluding true conductivity. As for Lorentz's claim that he could derive Fresnel's drag coefficient and other optical laws from his theory, Heaviside said he found the argument in the abstract too compressed to follow.

That Heaviside referred to Lorentz's charged particles as "electrification" is significant. It reflected his adherence to the strict Maxwellian view of electric charge as merely a surface manifestation of discontinuities in the field, without any independent or substantial existence. Bodies could be "electrified," Heaviside said, but there was no such thing as "electricity." The same extension of macroscopic concepts to the microscopic realm lay behind his remark in the preface to Volume I of his *Electromagnetic Theory*, written shortly after he had seen Lorentz's abstract, that Fresnel's drag hypothesis could best be understood by "supposing that the molecules of transparent matter act like little condensers in increasing the permittivity, and that the matter, when in motion, only carries forward the increased permittivity."[72] These "little condensers" would have much the same effect on passing light waves as would Lorentz's electrons, but conceptually the two ideas were quite different. Heaviside's suggestion was that of someone at home in the pages of the *Electrician*. Continental physicists had long drawn their models for the workings of ether and matter from physical astronomy, with its particles and forces; British physicists, from mechanical engineering, with its stresses and strains.[73] Heaviside drew instead on *electrical* engineering, and while "little condensers" were ubiquitous in electrical engineering, free charged particles like electrons were unknown.

71. H. A. Lorentz to Lodge, 8 Mar. 1891, OJL-UCL; Heaviside to Lodge, 13 Nov. 1893, OJL-UCL. Lorentz's abstract and Heaviside's comments are transcribed in Buchwald 1985a: 259–62; the words "and every dielectric polarization of its molecules" should be inserted after "such a medium" on p. 259, and "etherial" before "mag. energy" in Heaviside's comment at the bottom of p. 260.
72. Heaviside 1893–1912, vol. 1, "Preface" (unpaginated), [16 Dec. 1893]; cf. Heaviside to Lodge, 13 Nov. 1893, OJL-UCL, and Whittaker 1951: 403.
73. See Kargon 1969: 430.

Heaviside's roots in telegraphy led him to regard currents, fields, and electromagnetic waves as far more real and fundamental than charged particles and electrostatic forces. Earlier physicists had leaped quite naturally from their experience with electrostatic phenomena to the idea that currents were flows of the same electric fluids that exerted electrostatic forces. But Heaviside's own experience had been very different: "It so happened," he wrote in 1885,

> that my first acquaintance with electricity was with the dynamic phenomena, and after I had read with absorbed interest that instructive book, Tyndall's "Heat as a Mode of Motion." This may explain why, when it came later to book-learning regarding electricity, I had the greatest possible repugnance to all the explanations, and could not accept the electric current to be the motion of electricity (static) through a wire, but thought it something quite different.[74]

"I never swallowed the electric fluids," Heaviside later told FitzGerald; the idea of particles of "electricity" exerting forces on each other seemed irrelevant to the phenomena he had encountered in telegraphy.[75]

That a concern with electrical technology led to an emphasis on currents and fields at the expense of electrostatics was a commonplace among the Maxwellians. FitzGerald began a planned textbook on electrical engineering with the blunt statement, "To the Electrical Engineer current electricity is of so much greater importance than statical electricity that the study of the latter may be put aside as of secondary consequence and almost the whole attention be concentrated on the former," particularly the use of currents to convey and transform energy.[76] Lodge echoed him in the introduction to his 1906 book *Electrons,* quoting the remark of "our brilliant and lamented friend, G. F. FitzGerald," that electrostatics was "'one of the most beautiful and useless adaptations of nature'" and noting that a concern with "the practically useful" had led physics teachers "almost to ignore, or at any rate to scamper through, the domain of electrostatics, and to begin the study of electricity with the phenomena of *current,* especially with the connection between electricity and magnetism."[77] The predominance in the latter half of the nineteenth century of technologies based on currents and fields— first cable telegraphy, later dynamos and power systems—helped sus-

---

74. Heaviside 1892, 1: 435 [1885]; cf. Buchwald 1985c: 290.
75. Heaviside to FitzGerald, 13 Apr. 1894, FG-RDS.
76. FitzGerald manuscript 10380/27, FG-TCD Library. The four surviving pages of this draft are not part of the draft textbook on elementary physics with which they are filed.
77. Lodge 1906: xiv. Lodge said (p. xv) that his discussion of electrons was directed at the "comparatively leisured group" who were concerned not with practical problems but with "the gratuitous pursuit of philosophical speculation, exact experiment, and pure theory."

tain the field approach exemplified by Maxwellian theory and undercut any concern with forces between electrical particles.

Electron theory grew out of work on optics and cathode rays which was quite distant from Heaviside's own background and experience, and he always had mixed feelings about it. "I have much difficulty with the electron idea," he told FitzGerald in 1897. "In a sense, I believe in them. In fact, I recognised 10 years ago that to bring light phenomena into Maxwell's theory you want internal charges, *and in pairs*." But he was "rather prejudiced" against the idea that a conduction current was a real flow of electrons—it was, he thought, "more likely to be a local phenomenon in the main"—and he was puzzled by "what to understand by the material atom on which the charge (presumably) exists."[78] In Maxwellian theory, the discontinuities of displacement that constituted charge could exist only at the boundaries of conductors. But how could the existence of electrons explain conductivity, if they had themselves to be conductors? Lorentz had sidestepped this problem by simply positing the existence of charged particles in matter without specifying the nature of their "charge." Once this initial leap was made, the rest of his theory fit fairly well with Maxwellian principles, and when, "after profound research," Heaviside finally obtained a copy of Lorentz's 1895 *Versuch*, he had high praise for "this important application of Maxwell's theory to optical phenomena." (He also noted "with some personal gratification" Lorentz's use of vector analysis and of Heaviside's simplified field equations.)[79] But for all the strengths of Lorentz's theory, Heaviside could not regard it as really complete or fundamental as long as the nature of the material core of the electron was itself left unexplained.

Treating electrons as "singularities" in a rotational ether seemed to offer a way around this problem, and the prospect of a truly ultimate theory helped make Larmor's version of electron theory especially attractive to many British physicists in the late 1890s. Heaviside, however, would have none of it. "Larmor won't do," he told FitzGerald in 1897. "The rot[ational] ether won't work after Larmor any more than it did in 1891," when Heaviside had first explored its workings and detailed its failings.[80] Larmor had offered no satisfactory explanation of how his

78. Heaviside to FitzGerald, 12 July 1897, FG-RDS; Heaviside 1893–1912, 3: 50 [1900]. On electrons as "electrified *matter*," see Heaviside 1893–1912, 3: 98–99 [1901] and 3: 475 [1911]. J. J. Thomson viewed his "corpuscles" in a similar way; see Feffer 1989.

79. Heaviside 1893–1912, 3: 52–53 [1900]. Heaviside regretted that Lorentz's book had not been translated into English; "foreigners," he said, "seem to be gifted linguists quite naturally, so much so that they have invented a large number of lingos, and are commonly skilled in several at once," but Britons' lack of such facility meant they missed out on much valuable work.

80. Heaviside to FitzGerald, 12 July 1897, FG-RDS.

"strain-center" electrons remained self-locked, nor of how they exerted mechanical forces on one another, nor even of how they managed to move through the rotational ether, and Heaviside was irritated by what he saw as Larmor's attempts to hide these and other weak points of his theory behind clouds of convoluted verbiage and needlessly complicated mathematics. Larmor's failure to deliver on his original promise to explain the whole universe on a single dynamical basis was hardly a major offense, but "What I *should* like to see," Heaviside told FitzGerald in 1895, "would be a little more candour about the illsuccess of the rotational ether to satisfy electrical requirements."[81] Even when Larmor later tried to distance his theory from its roots in the rotational ether, he continued to treat his electrons as singularities in the ether rather than bits of electrified matter and to use analytical methods Heaviside regarded as antithetical to a truly Maxwellian approach. Heaviside and Larmor represented very different styles of physics, one rooted in electrical technology, the other in the mathematical traditions of Cambridge. It was perhaps inevitable that their views would diverge and that Heaviside would find himself out of sympathy with what he later called the "larmorial craze" that swept through British theoretical physics around the turn of the century.[82]

By the late 1890s, Heaviside was in even more danger than usual of drifting out of contact with the wider scientific community. FitzGerald and Lodge tried to prevent this and also to help him in more material ways. In 1896 they joined John Perry in a successful campaign to secure a Civil List pension of £120 a year for Heaviside—and what was perhaps more difficult, persuaded him to accept it.[83] Heaviside regarded this not as "charity" but as proper official recognition of his services to science, and though he joked about having been "pensioned off," he was pleased with the way it improved his standing with his family and neighbors at Paignton.[84] He was also happy to have the extra money, especially when the death of his father in November 1896 (his mother had died in 1894) led him to look for a new home away from his brother's music shop.

81. Ibid., 4 Feb. 1895. See Orr 1900 for a further critique of Larmor's rotational ether.

82. Heaviside's phrase "larmorial craze" was quoted back to him by Ludwig Silberstein in a letter of 10 Jan. 1917, OH-IEE. On Larmor's influence around 1900, see Warwick 1989. Heaviside's attitude toward Larmor was also colored by the knowledge that Larmor had sided with the mathematical "rigorists" who had blocked one of his papers from being published by the Royal Society in 1894; see Heaviside to FitzGerald, 4 Feb. 1895, FG-RDS, and Hunt 1991.

83. Perry to FitzGerald, 12 Feb. and 2 Mar. 1896, FG-RDS; Nahin 1988: 260. The pension was raised to £220 in 1914.

84. Heaviside to FitzGerald, 4 Mar. and 6 June 1896, FG-RDS, and 31 Aug. 1896, FG-TCD Physics; the latter includes a delightful mock conversation between "Mrs. Grundy" and "Mrs. Toady" about "Ollie" ("Before") and "Mr. Heaviside" ("After").

After a long search, he found and leased Bradley View, a house near Newton Abbot, a few miles north of Paignton. "Behold a transformation!" he wrote to FitzGerald in May 1897, shortly after moving in. "The man 'Ollie' of Paignton, who lives in the garrets at the music shop, is transformed into Mr. Heaviside, the gentleman who has taken Bradley View."[85]

But Heaviside's life at Bradley View soon turned sour. Neighborhood boys began to harass him, calling him names and, he claimed, "spying" on him. He got along no better with his adult neighbors; their minds were "perfectly vacuous," he said, adding a little inconsistently that "any space left is filled up with dirt." He felt completely isolated, he told Lodge in September 1898, and would "be glad to see an intelligent human being even for ½ hour."[86] Whether fortuitously or not, it was just two days later that FitzGerald arrived on what he called his "pilgrimage to the 'Sage of Devon.'" He spent a long Saturday afternoon at Bradley View, discussing deep issues of electromagnetic theory with Heaviside and then bicycling with him through the South Devon countryside.[87] Both men enjoyed the visit, but it gave Heaviside only a brief respite from his increasing isolation.

FitzGerald tried to bridge the gap between Heaviside and Larmor, but it had become a losing battle. Heaviside was "rather sarcastic" about Larmor's theory during FitzGerald's visit, and when the *Electrician* asked him to review *Aether and Matter* in May 1900, he declined, presumably to avoid having to write a negative review.[88] Instead, he bought a copy for himself—"a compliment I seldom pay to books," he told Larmor—and set about trying to reconcile its treatment of moving media with his own

85. Heaviside to FitzGerald, 23 May 1897, FG-RDS.
86. Heaviside to Lodge, 15 Sept. 1898, OJL-UCL. Heaviside was convinced that this harassment was a "put up job." He told Lodge that when a woman had earlier moved in nearby in Paignton, "I recognized her instantly as the woman who had spied upon me in London." Lodge apparently expressed some doubts, prompting Heaviside to reply (23 Sept. 1898, OJL-UCL), "Of course you don't understand—it took me some time before my detective instinct led me to understand. Perhaps you are too honest to even imagine the vile ways in which a crafty enemy can go to work to injure a solitary man." Less extreme indications of Heaviside's paranoia occur in his other letters to Lodge and FitzGerald from this period.
87. FitzGerald to Lodge, 18 Sept. 1898, OJL-UCL. FitzGerald made his "pilgrimage" after the Bristol meeting of the British Association, apparently on short notice and perhaps at Lodge's urging. The Cambridge physicist G. F. C. Searle also visited Heaviside at least five times in the 1890s and about once a year between 1905 and 1920; see Searle 1987.
88. FitzGerald to Lodge, 18 Sept. 1898, OJL-UCL; editor of *Electrician* to Heaviside, 10 May 1900, OH-IEE, offering him a premium rate of up to 30 shillings for a one- or two-column review. In the end, FitzGerald wrote the review; see FitzGerald 1902: 511–15 [1900].

OLIVER HEAVISIDE with his bicycle, early 1890s. Courtesy IEE
Archives

"Idiots consider me a madman about the bike; I ride every day."
—Oliver Heaviside to G. F. FitzGerald, May 1899

field-based approach.[89] He found several errors and apparent inconsistencies in Larmor's derivation of the Fresnel drag coefficient, and early in July he wrote to ask what FitzGerald thought of it.[90] FitzGerald was then visiting Larmor in Cambridge, and Heaviside's letter (now lost) was forwarded to him there. After discussing it with Larmor, FitzGerald wrote back that Larmor thought Heaviside had made the mistake of treating the "displacement" in the ether as a real displacement in space and so had compounded it directly with the polarization of matter by the separation of electrons—two things that had to be carefully distinguished in moving bodies, since it was only the material polarization that was carried along by the motion. FitzGerald agreed with Larmor: "This identification of the ether 'displacement' with an actual movement of the point of the ether in space" had long been a source of confusion, he told Heaviside, "and I think your difference with Larmor is due to your not distinguishing between the two."[91]

But Larmor and FitzGerald had blundered. Heaviside made it clear he had *never* believed that electric "displacement" represented a literal displacement in space, and FitzGerald soon apologized, telling Heaviside that he was "now sure that it is *not* on account of the point Larmor mentioned that you differ from him."[92] The real root of their difference lay deeper and involved Heaviside's failure to distinguish between the electric force on the ether that produced etherial displacement and the mechanical force on electrons owing to their motion through a magnetic field (the Lorentz force, $q\mathbf{v} \times \mathbf{B}$) that produced material electric polarization.[93] FitzGerald clarified this point in a series of letters to Heaviside in August and showed that when it was taken into account, and when the modified time derivative appropriate to moving media was used, Fresnel's drag coefficient followed quite directly. He also emphasized the fundamental asymmetry that was introduced into the electromagnetic

89. Heaviside to Larmor, 13 July 1900, JL-RS. In *Elec. 45* (1900): 637n, Heaviside recommended *Aether and Matter* "as a brain-stiffener during the hot weather." Larmor raised a mild objection (Larmor to Heaviside, 25 Aug. 1900, OH-IEE), and the remark was deleted when the section was reprinted in Heaviside 1893–1912, 3: 48.

90. Larmor later admitted that his derivation of the Fresnel drag in Larmor 1900: 60–61 was flawed, but he was able to correct it. See Larmor to Heaviside, 25 Aug. 1900, OH-IEE; Heaviside 1893–1912, 3: 59 [5 Oct. 1900]; and the sheet of "Addenda and Corrigenda" dated "Jan. 1901" in Larmor 1900.

91. FitzGerald to Heaviside, 12 July 1900, OH-IEE, quoted in Buchwald 1985a: 134. Letters to FitzGerald from this period have apparently all been lost.

92. FitzGerald to Heaviside, 8 Aug. 1900, OH-IEE. Heaviside later wrote that he had never heard of anyone mistaking Maxwell's "displacement" for a real spacial displacement "until one day I heard that I had been supposed to have done so myself"; Heaviside 1893–1912, 3: 98 [1901].

93. This is reflected in Heaviside's initial comments on Lorentz's force formula in 1893; see Heaviside to Lodge, 13 Nov. 1893, OJL-UCL, and Buchwald 1985a: 261–62.

equations by the fact that matter contained electrons but no corresponding "magnetons" (magnetic monopoles). As a result, for example, the motion of a body across an electric field produced no magnetic polarization, since there were no magnetons to be separated by a magnetic "Lorentz force." The microstructure of matter had to be taken explicitly into account when treating moving bodies, FitzGerald said; the old macroscopic and purely symmetrical approach broke down when applied in detail to such cases.[94] Indeed, he became convinced that the induction of currents by motion through a steady magnetic field could be explained *only* by assuming that conductors contain electrons. "Far from . . . being contrary to the spirit of Maxwell's treatise," he said in a review of *Aether and Matter,* the existence of electrons is "in reality quite compatible with it, and from some points of view even essential."[95] For FitzGerald, electrons had become an integral part of Maxwellian theory.

Heaviside was reluctant to take this final step, particularly since it would mean giving up the strict symmetry between electric and magnetic quantities that had been the foundation of his work since 1885. He tried hard to translate the results of electron theory into his own purely field-based idiom, and with some success: in the *Electrician* in October 1900 he showed that if, on general principles, we separate the electric displacement into etherial and material components, "we can eliminate the ionic hypothesis" and derive Fresnel's drag coefficient without invoking electrons. But if this separation into components were really independent of the "ionic hypothesis," it ought to apply to the magnetic flux as well as to the electric displacement. When the extra magnetic term was added on, however, the equations no longer yielded Fresnel's coefficient.[96] Heaviside had long accepted that optical evidence pointed to an underlying electrical structure of material media, whether pairs of electrons or his own "little condensers." But he drew a sharp line between such speculations about the constitution of matter and the abstract (and strictly symmetrical) framework of electromagnetic theory, which he believed should not be changed lightly, if at all. "I think it will be wise not to be overhasty in rejecting terms in the circuital equations because magnetons do not exist," he wrote in the *Electrician.* "Let us try to keep to symmetry to aid understanding any want of symmetry that may be forced upon us by special hypotheses."[97] In 1900, Heaviside still regarded the

94. FitzGerald to Heaviside, 8 and 12 Aug. 1900, OH-IEE.
95. FitzGerald 1902: 512 [1900]; cf. FitzGerald to Larmor, 11 July 1900, JL-RS. On the problem of induction by motion, see Knudsen 1980 and Miller 1981: 150–61.
96. Heaviside 1893–1912, 3: 55 [5 Oct. 1900].
97. See the discussion of moving media in Heaviside 1893–1912, 3: 44–52 [17 Aug. 1900] and 3: 52–61 [5 Oct. 1900]; the quotation is from p. 58.

existence of electrons as a "special hypothesis." But to FitzGerald—and to Lorentz, Larmor, and a whole generation of younger physicists—the existence of electrons (and the nonexistence of magnetons) was by then an established fact and the proper starting point for any analysis of material media. They regarded the electron as both the capstone of the Maxwellian program and the first step toward a new physics of the microstructure of matter.

Conclusion

Toward the end of his 1906 book *Electrons,* Lodge noted that "the electric current is a more material entity, or has a more nearly material aspect, than was thought probable a little while since"; his *Modern Views,* with its stout denial that anything really flowed in an electric current, no longer seemed quite so modern. But all he had said before about the way currents are driven by the surrounding field, "all that was taught about the paths by which the energy travels and arrives at point after point of the conductor, there to be dissipated as heat"—all of this, he insisted, remained as true as ever.[98] The advent of the electron had transformed the way charges and currents were conceived, but the Maxwellians had always regarded these as somewhat secondary issues. The core of Maxwellian theory was its account of the propagation of waves and energy through the field, and this, Lodge said, had been left untouched.

Maxwellian theory was a product of its time and place. It embodied a set of concerns—above all with energy and its propagation—characteristic of late Victorian Britain. Heaviside's work on telegraphic propagation, FitzGerald's on electromagnetic waves, and Lodge's on ether models all reflected an interest in localizing the energy in the field and making it manifest. The system of "absolute" electrical units promulgated by the British Association, and the emphasis on precision measurement that went with it, were also based on a concern with energy and the ways in which it could be conveyed and put to use. Most tellingly of all, "Maxwell's equations" had themselves been recast to make them better express the localization and flow of energy in the field. Heaviside's abandonment of the vector and scalar potentials in favor of the electric and magnetic forces was part of a deliberate effort to make the equations reflect the true state of the field, conceived in terms of the energy at each point.

---

98. Lodge 1906: 198. Lodge reinforced this point by issuing *Modern Views,* 3d ed., in 1907.

From the mid-1890s, British electrical physics underwent a major shift as the influence of Continental traditions and new experimental work combined to focus increasing attention on the microscopic structure of matter, centering initially on the electron. Maxwellian theory was modified and in a sense superseded by the advent of electron theory, as physicists began to ask new questions and to offer new answers in which the old macroscopic approach was treated as merely approximate. But Maxwellian theory was not abandoned; indeed, its most important parts, its treatment of free fields, energy flux, and electromagnetic waves, were transferred bodily into the new physics of the twentieth century and applied more widely than ever in new areas of both science and technology. The Maxwellians' notions of charge and current were replaced by a new picture based on the electron, but the field equations they had devised in the 1880s continued to occupy a central place in electromagnetic theory and also to reflect the concern with energy and its propagation that had given unity and purpose to their work.

Epilogue

The Maxwellian group broke up rather abruptly within a year or two after the turn of the century. The death of Hertz in January 1894 had been a serious loss, but it was partly compensated by the closer contacts with Larmor that began in 1893, and the Maxwellians were able to carry on as a reasonably close-knit group for the remainder of the decade, despite the undeniable friction between Heaviside and Larmor. After 1900, however, a combination of death, illness, and career changes conspired to split up the group.

Larmor's career progressed well in these years but underwent little dramatic change; he continued to lecture at St. John's College, Cambridge, and in 1903 was chosen to succeed Stokes as Lucasian Professor. Knighthood in 1909 merely put the cap on his past achievements; he produced little creative new work after the publication of *Aether and Matter* in 1900. In his later years he acquired the reputation—apparently carefully cultivated—of being something of a curmudgeon. He served as a Conservative member of Parliament for Cambridge University from 1911 to 1922 and became as conservative in science, steadfastly opposing the new relativity and quantum theories, as he was in politics. Larmor was reportedly fond of saying that "all true scientific progress ceased around 1900," and this seems to have indeed been largely true of his own work.[1]

Lodge's creative scientific work, too, essentially came to an end in 1900. After achieving great success and acclaim for his whirling-ma-

---

1. Eddington 1942: 205.

chine experiments and his pioneering work on wireless telegraphy in the 1890s, he was offered the principalship of the new University of Birmingham by Joseph Chamberlain in the spring of 1900. He hesitated, knowing that taking the post would effectively end his research career, but finally decided to accept. Lodge served as principal until 1919, becoming increasingly prominent as a public figure (he was knighted in 1902) and as a spokesman for the reform of English higher education. He was perhaps best known for his psychical researches and his attempts to communicate with the dead. But his work in physics was limited to writing popular and semipopular books, notably *Electrons* (1906), *The Ether of Space* (1909), and *Ether and Reality* (1925); and though he lived until 1940, Lodge made few contributions to the new physics of the twentieth century.[2]

Heaviside's work also went into sharp decline after 1900, largely because of failing health. His 1897 move to Newton Abbot, a few miles inland from Paignton, led to many troubles. Being, as he told FitzGerald, "a thin skinned fatless man, very sensitive to cold," Heaviside was worried that the somewhat harsher climate of his new home might affect his health, and it apparently did.[3] He suffered from jaundice in the winter of 1901–2 and was ill again in 1906; in 1908, severe illness and increasing harassment by neighborhood boys who threw stones at his house ("Panes [were] broken and splashed over my sickbed," he said) forced him to move a few miles away to Torquay, where he lodged with his brother's sister-in-law. He slowly recovered his health but was never again entirely well, and in the years up to his death in 1925 he became more and more solitary and eccentric.[4]

These bouts of illness effectively ended Heaviside's scientific work. The *Electrician* stopped carrying his regular series "Electromagnetic Theory" in 1902, and over the next few years he published little more than some short notes in *Nature* on electromagnetic waves and moving charges. He eventually managed to fill out the third volume of his *Electromagnetic Theory*, published in 1912, but its later sections contained little that was really new.[5] After 1902, or 1906 at the latest, Heaviside was not an active member of the scientific community.

By far the greatest blow to the Maxwellian group was the death of

2. On this part of Lodge's career, see Lodge 1931b: 314–21, Jolly 1974: 131–238, and Goldberg 1970.

3. Heaviside to FitzGerald, 23 May 1897, FG-RDS.

4. Heaviside to Lodge, 10 Dec. 1908, OJL-UCL; see also Searle 1987 and Nahin 1988: 284–97.

5. Claims that Heaviside made remarkable new discoveries in the 1910s and 1920s are decisively refuted in Gossick 1977.

FitzGerald in February 1901. He had suffered for years from recurring digestive problems, apparently caused by an ulcer, but had always seemed to recover well. Then toward the end of August 1900 he was struck by what he described more than four months later as "this last long attack of indigestion which has never been really well since then." It had started while he was preparing a discussion paper on "ions" for the British Association meeting. "I don't like to think it," he wrote to Lodge in January 1901, "but I am rather afraid these attacks begin with questions that run in my head for a week or so. If I have to give up this continuous thinking I am afraid I cannot do much more in the rest of my life: but I hope I may be wrong in my hypothesis. Anyway just now I am under doctor's orders not to think at all."[6] He then went on, however, to fill more than a page with acute remarks about the vortex ether, the paths of energy flux, and his latest ideas about the nature of conduction currents.

Lodge was alarmed by FitzGerald's letter and immediately wrote to Larmor, Heaviside, and others to ask them not to trouble FitzGerald with letters until he could rest and recover.[7] Late in January, FitzGerald began a "rest cure" of seclusion and massage, but his condition did not improve. An operation was finally attempted on 22 February 1901 and was reportedly successful, but FitzGerald's strength had been sapped and he died a few hours later.[8] He was forty-nine. He left a widow and eight young children.

Although Heaviside had been told that FitzGerald was ill, he said later that "a postcard from Perry, FitzGerald is dead! came as a great shock, and I am not easily shocked." Their friendship, though conducted almost entirely through the post, had become both deep and strong; "I came to love the man," Heaviside said, and he regarded FitzGerald's premature death as "a national misfortune."[9] For Heaviside and Larmor, and especially for Lodge, it was also a profound personal loss. Looking back on his life in 1931, Lodge wrote in his autobiography that his two closest friends had been F. W. H. Myers, his colleague in psychical research, and FitzGerald. Myers died in January 1901 and FitzGerald scarcely a month later. "He and Myers were like brothers to me," Lodge told Rayleigh a few days after FitzGerald's death. "It makes one feel lonely."[10]

6. FitzGerald to Lodge, 16 Jan. 1901, filed among Larmor's letters to Lodge, OJL-UCL.

7. Lodge to Larmor, 18 Jan. 1901, JL-RS; Heaviside to Lodge, 19 Jan. 1901, OJL-UCL.

8. E. P. Culverwell to Larmor, 19 Feb. 1901, JL-RS; Larmor to Lodge, 23 Feb. 1901, OJL-UCL.

9. Heaviside to Lodge, 27 Feb. 1901, OJL-UCL.

10. Lodge 1931b: 220; Lodge to Rayleigh, 24 Feb. 1901, Rayleigh-AIP.

FitzGerald performed no great original experiments like those of Hertz and Lodge, nor did he ever write a book as widely read as Lodge's *Modern Views* or as profound and influential as Larmor's *Aether and Matter* or Heaviside's *Electromagnetic Theory*. Indeed, he never wrote a book at all. But the short papers he did write—"not large in bulk," as Heaviside said, "but very choice and original"—were of unusual value and often pointed the way to important advances, as in his papers on the generation of electromagnetic waves and on the field produced by a moving charge.[11] FitzGerald exerted his strongest influence, however, not through his published papers but through his personal contacts with other scientists. He was "the life and soul of debate" at British Association meetings, Lodge said, and was by all accounts extraordinarily free with his time in answering the queries of others and in commenting on their work.[12] As a result, many of his best and most fruitful ideas were submerged in the writings of others; his contraction hypothesis, for example, became known only through Lodge's papers, and his crucial contribution to the genesis of Larmor's electron theory was almost completely unknown to other physicists at the time.

FitzGerald seems to have made little effort to win credit for himself. His real aim was to promote scientific progress along the lines he thought important, and he found that the most effective way to do this, the way most suited to his particular talents and personality, was to toss out suggestions and seek to stimulate the work of others. He did this among his own colleagues at Dublin; he did it among the broader community of British physicists; and most important, he did it with Lodge, Heaviside, Hertz, and Larmor, helping to hold them together as a group and stimulating their continuing efforts to clarify and extend Maxwell's theory. A large part of their collective achievement can be traced to FitzGerald's influence. He was in many ways the soul of the Maxwellian group. With his death, it came to an end.

11. Heaviside to Perry, [Feb. 1901], quoted in Larmor 1901a, in FitzGerald 1902: xxvi.
12. Lodge 1901, in FitzGerald 1902: xx.

From Maxwell's Equations
to "Maxwell's Equations"

The set of four vector equations now generally known as "Maxwell's equations" does not appear in Maxwell's *Treatise*. The chapter called the "General Equations of the Electromagnetic Field" instead contains a list of thirteen principal equations, lettered (A) through (L), in which the vector and scalar potentials figure prominently. Heaviside found this long list unwieldy and argued that a more compact set of equations involving only the electric and magnetic forces and fluxes would be clearer and more useful, particularly in the treatment of electromagnetic propagation and energy flow. He devised what are now known as "Maxwell's equations" in 1884 and published them in the *Electrician* early in 1885; by the mid-1890s they had won acceptance as the standard form of Maxwell's theory.

Maxwell stated the "General Equations" in his *Treatise* first in full Cartesian coordinates and then more compactly in quaternions (indicated by Gothic characters). Heaviside combined and modified eight of these to obtain his four basic equations. Both he and Maxwell used the prefix "V" to indicate a vector product; Maxwell also used "S" to indicate a scalar product.

	Maxwell		*Heaviside*
	Cartesians	*Quaternions*	*Vectors*
(A)	$a = dH/dy - dG/dz$ $b = dF/dz - dH/dx$ $c = dG/dx - dF/dy$	$\mathfrak{B} = V\nabla\mathfrak{A}$	$\mathbf{B} = \mathrm{curl}\ \mathbf{A}$
(A')	$da/dx + db/dy + dc/dz = 0$	$S\nabla\mathfrak{B} = 0$	$\mathrm{div}\ \mathbf{B} = 0$

(B) $P = cdy/dt - bdz/dt -$ $\mathfrak{E} = Vu\mathfrak{B} - \dot{\mathfrak{A}} - \nabla\psi$ $\mathbf{E} = Vu\mathbf{B} - \dot{\mathbf{A}} - \text{grad } \psi$
      $dF/dt - d\psi/dx$
    $Q = adz/dt - cdx/dt -$
      $dG/dt - d\psi/dy$
    $R = bdx/dt - ady/dt -$
      $dH/dt - d\psi/dz$

(E) $4\pi u = d\gamma/dy - d\beta/dz$ $4\pi\mathfrak{E} = V\nabla\mathfrak{H}$ $\mathbf{C} = \text{curl } \mathbf{H}$
    $4\pi v = d\alpha/dz - d\gamma/dx$
    $4\pi w = d\beta/dx - d\alpha/dy$

(F) $f = K/4\pi\, P$ $\mathfrak{D} = K/4\pi\ \mathfrak{E}$ $\mathbf{D} = \epsilon\mathbf{E}$
    $g = K/4\pi\, Q$
    $h = K/4\pi\, R$

(I) $u = CP + K/4\pi dP/dt$ $\mathfrak{C} = (C + K/4\pi d/dt)\mathfrak{E}$ $\mathbf{C} = (k + \epsilon d/dt)\mathbf{E}$
    $v = CQ + K/4\pi dQ/dt$
    $w = CR + K/4\pi dR/dt$

(J) $df/dx + dg/dy + dh/dz = \rho$ $S\nabla\mathfrak{D} = \rho$ $\text{div } \mathbf{D} = \rho$

(L) $a = \mu\alpha$ $\mathfrak{B} = \mu\mathfrak{H}$ $\mathbf{B} = \mu\mathbf{H}$
    $b = \mu\beta$
    $c = \mu\gamma$

Heaviside combined (A′) and (L) to obtain div ($\mu\mathbf{H}$) = 0; (J) and (F) to obtain div ($\epsilon\mathbf{E}$) = ρ; and (E) and (I) to obtain curl $\mathbf{H} = k\mathbf{E} + \epsilon\dot{\mathbf{E}}$, which he called the "first circuital equation." His next step was less obvious and more important. His discovery of the energy flux formula in 1884 convinced him that the first circuital equation should be matched by a second one relating the curl of the electric force to the rate of change of the magnetic flux. By restricting (B) to the case of a stationary medium, so that $\mathbf{u} = 0$ (Heaviside took the effect of motion into account separately, as an "impressed force"), he obtained:

$$\mathbf{E} = -\dot{\mathbf{A}} - \text{grad } \psi.$$

Taking the curl of both sides then yielded:

$$\text{curl } \mathbf{E} = \text{curl } (-\dot{\mathbf{A}}) - \text{curl } (\text{grad } \psi).$$

But the curl of the gradient of any function is zero, so the second term on the right vanishes. Interchanging the order of the time and space differentiations of $\mathbf{A}$ then yielded:

$$\text{curl } \mathbf{E} = -d/dt \,(\text{curl } \mathbf{A}).$$

Finally, combining this with (A) and (L) yielded:

$$-\text{curl } \mathbf{E} = \mu\dot{\mathbf{H}}.$$

This was Heaviside's "second circuital equation." It allowed him to dispense with the scalar and vector potentials and to express Maxwell's theory in a very compact form:

$$\text{div } \epsilon\mathbf{E} = \rho \qquad \text{curl } \mathbf{H} = k\mathbf{E} + \epsilon\dot{\mathbf{E}}$$
$$\text{div } \mu\mathbf{H} = 0 \qquad -\text{curl } \mathbf{E} = \mu\dot{\mathbf{H}}.$$

He then made the equations more general and symmetrical by introducing hypothetical magnetic charge (σ) and conductivity (g) and by adding terms to represent convection currents ($\mathbf{u}\rho$ and $\mathbf{u}\sigma$) and impressed forces ($\mathbf{h}_0$ and $\mathbf{e}_0$), particularly the "motional" impressed forces $\mathbf{h} = V\epsilon\mathbf{E}\mathbf{u}$ and $\mathbf{e} = -V\mu\mathbf{H}\mathbf{u}$. This yielded the dynamically complete set discussed in Heaviside 1892, 2: 539–41:

$$\text{div}\epsilon\mathbf{E} = \rho \qquad \text{curl } (\mathbf{H} - \mathbf{h}_0 - \mathbf{h}) = k\mathbf{E} + \epsilon\dot{\mathbf{E}} + \mathbf{u}\rho$$
$$\text{div}\mu\mathbf{H} = \sigma \qquad -\text{curl } (\mathbf{E} - \mathbf{e}_0 - \mathbf{e}) = g\mathbf{H} + \mu\dot{\mathbf{H}} + \mathbf{u}\sigma.$$

Abbreviations

The following abbreviations are used in the notes and bibliography.

BOOKS AND JOURNALS

AHES	*Archive for History of the Exact Sciences*
Ann. Sci.	*Annals of Science*
BA Report	*Report of the British Association for the Advancement of Science*
BJHS	*British Journal for the History of Science*
DNB	*Dictionary of National Biography*
DSB	*Dictionary of Scientific Biography*
Elec.	*The Electrician*
HSPS	*Historical Studies in the Physical [and Biological] Sciences*
JIEE	*Journal of the Institution of Electrical Engineers*
JSTEE	*Journal of the Society of Telegraph Engineers and Electricians*
Phil. Mag.	*Philosophical Magazine*
Phil. Trans.	*Philosophical Transactions of the Royal Society of London*
Proc. Phys. Soc.	*Proceedings of the Physical Society of London*
Proc. RS	*Proceedings of the Royal Society of London*
Sci. Proc. RDS	*Scientific Proceedings of the Royal Dublin Society*

Bibliography

MANUSCRIPT COLLECTIONS

G. F. FitzGerald Collection. Royal Dublin Society.
    Most of the surviving correspondence of G. F. FitzGerald is held by the Royal
    Dublin Society. The collection includes about ninety letters from Lodge, forty-five
    from Heaviside, and sixty from Larmor, along with several hundred from other
    scientists.
———. Trinity College Dublin Library.
    Eight of FitzGerald's research notebooks and some related materials are held in
    the manuscripts room of the main library.
———. Trinity College Dublin Physics Department.
    Much of FitzGerald's personal library, including his annotated copy of Maxwell's
    Treatise, is held by the FitzGerald Library of the Physics Department, along with
    some related manuscript materials.
Oliver Heaviside Collection. Institution of Electrical Engineers.
    A large collection of Oliver Heaviside's papers and correspondence is held by the
    Institution of Electrical Engineers in London. Most of the letters from Lodge and
    Hertz have been lost, but about sixty letters from FitzGerald, eighteen from Lar-
    mor, and hundreds from other scientists and engineers survive, as do Heaviside's
    research notebooks. Microfilm copies of most of this collection are held by the
    Niels Bohr Library of the American Institute of Physics in New York.
Heinrich Hertz Collection. Deutsches Museum.
    The surviving scientific correspondence of Heinrich Hertz is held by the Deut-
    sches Museum in Munich. I did not visit Munich myself, but the archivists at the
    museum kindly sent me photocopies of nine letters from Heaviside, five from
    FitzGerald, and three from Lodge. These, along with Hertz's replies (where avail-
    able) and related documents, have now been published in O'Hara and Pricha
    1987.
Joseph Larmor Collection. Royal Society.
    Most of the surviving scientific correspondence of Joseph Larmor is held by the

Royal Society of London. The collection includes about seventy-five letters from FitzGerald, one hundred from Lodge, and eight from Heaviside.

Oliver Lodge Collection. University College London.
The bulk of Oliver Lodge's surviving scientific correspondence is held by University College London Library. The collection (Add. MS. 89) includes about 185 letters from FitzGerald, 135 from Heaviside, and 160 from Larmor.

———. University of Birmingham.
A large collection of Lodge's nonscientific correspondence, along with a few letters from scientists, is held in the university's main library.

———. University of Liverpool.
Many of Lodge's research notebooks are held by the university library.

J. C. Maxwell Collection. University Library, Cambridge.
Most of the surviving correspondence of James Clerk Maxwell (Add. MS. 7655) is held by the University Library at Cambridge.

Rayleigh Collection. American Institute of Physics.
Microfilm copies of the correspondence of Lord Rayleigh [J. W. Strutt] are held by the Niels Bohr Library of the American Institute of Physics in New York. The originals are at the Air Force Geophysics Library, Hanscom AFB, Mass.

Henry A. Rowland Collection. Johns Hopkins University.
The scientific correspondence of H. A. Rowland (MS. 6) is held by the Milton S. Eisenhower Library at Johns Hopkins University.

Royal Society Archives.
The Royal Society of London retains in its archives most of the referee reports on papers submitted for publication in its *Philosophical Transactions*.

G. G. Stokes Collection. University Library, Cambridge.
The surviving correspondence of G. G. Stokes (Add. MS. 7656) is held by the University Library at Cambridge. Much of it concerns his work as secretary of the Royal Society.

J. J. Thomson Collection. University Library, Cambridge.
Most of the surviving correspondence of J. J. Thomson (Add. MS. 7654) is held by the University Library at Cambridge, including several interesting letters from FitzGerald.

William Thomson [Kelvin] Collection. University Library, Cambridge.
Much of the surviving correspondence of William Thomson [Lord Kelvin] (Add. MS. 7342) is held by the University Library at Cambridge, including about twenty letters from FitzGerald and several from Heaviside, Lodge, and Larmor.

BOOKS AND ARTICLES

Ahvenainen, Jorma. 1981. *The Far Eastern Telegraphs: The History of Telegraphic Communications between the Far East, Europe and America before the First World War*. Helsinki: Suomalainen Tiedeakatemia.

Appleyard, Rollo. 1930. *Pioneers of Electrical Communication*. London: Macmillan.

———. 1939. *The History of the Institution of Electrical Engineers*. London: Institution of Electrical Engineers Press.

Baker, E. C. 1976. *Sir William Henry Preece, F.R.S., Victorian Engineer Extraordinary*. London: Hutchinson.

Barty-King, Hugh. 1979. *Girdle Round the Earth: The Story of Cable and Wireless and Its Predecessors to Mark the Group's Jubilee, 1929–1979*. London: Heinemann.

Besterman, Theodore. 1935. *A Bibliography of Sir Oliver Lodge, F.R.S.* Oxford: Oxford University Press.

[Biggs, C. H. W.] 1889. "The Presidential Address." *Electrical Engineer 3:*52–53.

Bork, A. M. 1966a. "Maxwell and the Electromagnetic Wave Equation." *American Journal of Physics 35:*844–49.

———. 1966b. "The 'FitzGerald' Contraction." *Isis 57:*199–207.

———. 1967. "Maxwell and the Vector Potential." *Isis 58:*210–22.

Bowers, Brian. 1975. *Sir Charles Wheatstone.* London: Her Majesty's Stationery Office.

Bright, Charles. 1898. *Submarine Telegraphs.* London: Crosby Lockwood. Repr. New York: Arno, 1974.

Brittain, James E. 1970. "The Introduction of the Loading Coil: George A. Campbell and Michael Pupin." *Technology and Culture 11:*36–57.

Brush, S. G. 1967. "Note on the History of the FitzGerald-Lorentz Contraction." *Isis 58:*230–32.

Buchwald, Jed Z. 1985a. *From Maxwell to Microphysics: Aspects of Electromagnetic Theory in the Last Quarter of the Nineteenth Century.* Chicago: University of Chicago Press.

———. 1985b. "Modifying the Continuum: Methods of Maxwellian Electrodynamics." In P. M. Harman, ed., *Wranglers and Physicists: Studies on Cambridge Physics in the Nineteenth Century,* pp. 225–41. Manchester: Manchester University Press.

———. 1985c. "Oliver Heaviside: Maxwell's Apostle and Maxwellian Apostate." *Centaurus 28:*288–330.

Campbell, Lewis, and William Garnett. 1882. *The Life of James Clerk Maxwell.* London: Macmillan. Repr. Ed. Robert H. Kargon. New York: Johnson Reprint, 1969.

Cardwell, D. S. L. 1972. *The Organization of Science in England.* 2d ed. London: Heinemann.

Chalmers, A. F. 1974. "The Limitations of Maxwell's Electromagnetic Theory." *Isis 65:*469–83.

Chrystal, George. 1882. "Clerk Maxwell's 'Electricity and Magnetism.'" *Nature 25:*237–40.

Coates, Vary, and Bernard Finn. 1979. *A Retrospective Technology Assessment: Submarine Telegraphy—The Transatlantic Cable of 1866.* San Francisco: San Francisco Press.

Crowe, Michael J. 1967. *A History of Vector Analysis.* Notre Dame, Ind.: University of Notre Dame Press. Repr. New York: Dover, 1985.

Culley, R. S. 1871. *Handbook of Practical Telegraphy.* 5th ed. London: Longmans.

D'Agostino, Salvo. 1975. "Hertz's Researches on Electromagnetic Waves." *HSPS 6:*261–323.

De Tunzelmann, G. W. 1888. "Hertz's Researches on Electrical Oscillations." *Elec. 21* and *22.* Repr. *JSTEE 17:*717–59.

Dubbey, J. M. 1963. "The Introduction of the Differential Notation to Great Britain." *Ann. Sci. 19:*37–48.

Duhem, Pierre. 1954. *The Aim and Structure of Physical Theory.* Trans. P. P. Wiener from 2d French ed., 1914. Princeton, N.J.: Princeton University Press.

Eddington, A. S. 1942. "Sir Joseph Larmor." *Obituary Notices of Fellows of the Royal Society 4:*197–207.

Einstein, Albert. 1905. "Zur Elektrodynamik bewegter Körper." *Annalen der Physik 17:*891–921. Trans. in Miller 1981, pp. 392–415.

Everitt, C. W. F. 1975. *James Clerk Maxwell, Physicist and Natural Philosopher.* New York: Scribner's.

Faraday, Michael. 1855. *Experimental Researches in Electricity.* Vol. 3. London: Taylor and Francis. Repr. New York: Dover, 1965.

Feffer, Stuart M. 1989. "Arthur Schuster, J. J. Thomson, and the Discovery of the Electron." *HSPS* 20:33–61.

Feynman, Richard. 1964. *The Feynman Lectures.* 3 vols. Reading, Mass.: Addison-Wesley.

[FitzGerald, G. F.] 1881. "Works of James MacCullagh." *Nature 24:*26–27.

FitzGerald, G. F. 1885a. "On a Model Illustrating Some Properties of the Ether." Abstract. *Nature 31:*499.

———. 1885b. "Molecular Dynamics." *Nature 31:*503.

———. 1889. "The Ether and the Earth's Atmosphere." *Science 13:*390.

———. 1890. "On an Episode in the Life of J (Hertz's Solution of Maxwell's Equations)." *BA Report,* pp. 755–57.

[———.] 1891. "Hertz's Experiments." *Nature 43:*536–38, *44:*12–14, 31–35.

[———.] 1893. "The Propagation of Electric Energy." Review of Hertz, *Untersuchungen über die Ausbreitung der elektrischen Kraft. Nature 48:*538–39.

[———.] 1894. Review of Heaviside, *Electromagnetic Theory,* vol. 1. *Elec. 33:*104–6.

[———.] 1898a. "Notes" [on Larmor]. *Elec. 41:*239–40.

———. 1898b. "The Zeeman Effect and Disperson." *Science Progress 7:*416–29.

———. 1902. *The Scientific Writings of the Late George Francis FitzGerald.* Ed. Joseph Larmor. Dublin: Hodges and Figgis.

Föppl, August. 1894. *Einfuhrung in die Maxwell'sche Theorie der Elektricität.* Leipzig: Teubner.

Forbes, George. 1885. "Molecular Dynamics." Report on William Thomson's Baltimore lectures. *Nature 31:*461–63, 508–10, 601–3.

Friedman, Robert Marc. 1989. *Appropriating the Weather: Vilhelm Bjerknes and the Construction of a Modern Meteorology.* Ithaca, N.Y.: Cornell University Press.

Glazebrook, R. T. 1881. "On the Molecular Vortex Theory of Electromagnetic Action." *Phil. Mag. 11:*397–413.

———. 1928. "H. A. Lorentz." *Nature 121:*287–88.

Goldberg, Stanley. 1970. "In Defense of Ether: The British Response to Einstein's Theory of Relativity, 1905–1911." *HSPS 2:*89–125.

Gooday, Graeme. 1989. "Precision Measurement and the Genesis of Physics Teaching Laboratories in Victorian Britain." Ph.D. diss., University of Kent at Canterbury.

Gooding, David. 1980. "Faraday, Thomson, and the Concept of the Magnetic Field." *BJHS 13:*91–120.

Gossick, B. R. 1976. "Heaviside and Kelvin: A Study in Contrasts." *Ann. Sci. 33:*275–87.

———. 1977. "Where Is Heaviside's Manuscript for Volume 4 of His *Electromagnetic Theory?*" *Ann. Sci. 34:*601–6.

Hankins, Thomas L. 1980. *Sir William Rowan Hamilton.* Baltimore: Johns Hopkins University Press.

Headrick, Daniel. 1988. *Tentacles of Progress: Technology Transfer in the Age of Imperialism, 1850–1940.* Oxford: Oxford University Press.

Heaviside, Oliver. 1889. *Electromagnetic Waves.* London: privately printed.

———. 1892. *Electrical Papers.* 2 vols. London: Macmillan. Repr. New York: Chelsea, 1971.

———. 1893–1912. *Electromagnetic Theory.* 3 vols. London: Electrician Co. Repr. New York: Chelsea, 1971.

Heimann, P. M. 1971. "Maxwell, Hertz, and the Nature of Electricity." *Isis 62:*149–57.

Hendry, John. 1983. "Monopoles before Dirac." *Studies in History and Philosophy of Science 14:*81–87.

Hertz, Heinrich. 1892. *Untersuchungen über die Ausbreitung der elektrischen Kraft.* Leipzig: Barth.

——. 1893. *Electric Waves: Being Researches on the Propagation of Electric Action with Finite Velocity through Space.* Trans. D. E. Jones from 1st German ed., 1892. London: Macmillan. Repr. New York: Dover, 1962.

——. 1896. *Miscellaneous Papers.* Trans. D. E. Jones and G. A. Schott from 1st German ed., 1895. London: Macmillan.

——. 1899. *The Principles of Mechanics Presented in a New Form.* Trans. D. E. Jones and J. T. Walley from 1st German ed., 1894. London: Macmillan. Repr. New York: Dover, 1956.

——. 1977. *Memoirs, Letters, Diaries.* 2d ed. Ed. Johanna Hertz. Trans. Lisa Brinner, Mathilde Hertz, and Charles Susskind. San Francisco: San Francisco Press.

Hicks, W. M. 1885. "On the Constitution of the Luminiferous Ether on the Vortex Atom Theory." *BA Report,* p. 930.

Holmes, N. J. 1871. "Submarine Telegraphs." *Nature 4:*8–10.

Hughes, Thomas P. 1983. *Networks of Power: Electrification in Western Society, 1880–1930.* Baltimore: Johns Hopkins University Press.

Hunt, Bruce J. 1983. "'Practice vs. Theory': The British Electrical Debate, 1888–1891." *Isis 74:*341–55.

——. 1986. "Experimenting on the Ether: Oliver J. Lodge and the Great Whirling Machine." *HSPS 16:*111–34.

——. 1987. "'How My Model Was Right': G. F. FitzGerald and the Reform of Maxwell's Theory." In Robert Kargon and Peter Achinstein, eds., *Kelvin's Baltimore Lectures and Modern Theoretical Physics: Historical and Philosophical Perspectives,* pp. 299–321. Cambridge, Mass.: MIT Press.

——. 1988. "The Origins of the FitzGerald Contraction." *BJHS 21:*67–76.

——. 1991. "Rigorous Discipline: Oliver Heaviside versus the Mathematicians." In Peter Dear, ed., *The Literary Structure of Scientific Argument: Historical Studies,* pp. 72–95. Philadelphia: University of Pennsylvania Press.

——. Forthcoming. "Michael Faraday, Cable Telegraphy, and the Rise of Field Theory." *History of Technology.*

Jenkin, Fleeming, ed. 1873. *Reports of the Committee on Electrical Standards Appointed by the British Association for the Advancement of Science.* London: Spon.

Jolly, W. P. 1974. *Sir Oliver Lodge.* London: Constable.

Jordan, D. W. 1982a. "The Adoption of Self-Induction by Telephony, 1886–1889." *Ann. Sci. 39:*433–61.

——. 1982b. "D. E. Hughes, Self-Induction, and the Skin-Effect." *Centaurus 26:*123–53.

Jungnickel, Christa, and Russell McCormmach. 1986. *Intellectual Mastery of Nature: Theoretical Physics from Ohm to Einstein.* 2 vols. Chicago: University of Chicago Press.

Kargon, Robert. 1969. "Model and Analogy in Victorian Science: Maxwell's Critique of the French Physicists." *Journal of the History of Ideas 30:*423–36.

Kargon, Robert, and Peter Achinstein, eds. 1987. *Kelvin's Baltimore Lectures and Modern Theoretical Physics: Historical and Philosophical Perspectives.* Cambridge, Mass.: MIT Press.

Kelvin. See Thomson, William.

Kerr, John. 1876. "On the Rotation of the Plane of Polarization by Reflection from the Pole of a Magnet." *BA Report,* pp. 40–41.

——. 1877. "On the Rotation of the Plane of Polarization by Reflection from the Pole of a Magnet." *Phil. Mag. 3:*321–43.

Klein, Martin J. 1972. "Mechanical Explanation at the End of the Nineteenth Century." *Centaurus 17:*58–82.

Knott, C. G. 1911. *Life and Scientific Work of Peter Guthrie Tait.* Cambridge: Cambridge University Press.

Knudsen, Ole. 1976. "The Faraday Effect and Physical Theory, 1845–1873." *AHES 15:*235–81.

———. 1980. "19th Century Views on Induction in Moving Conductors." *Centaurus 24:*346–60.

———. 1985. "Mathematics and Physical Reality in William Thomson's Electromagnetic Theory." In P. M. Harman, ed., *Wranglers and Physicists: Studies on Cambridge Physics in the Nineteenth Century,* pp. 149–79. Manchester: Manchester University Press.

Kohlstedt, Sally Gregory. 1980. "*Science:* The Struggle for Survival, 1880 to 1894." *Science 209:*33–42.

Larmor, Joseph. 1900. *Aether and Matter.* Cambridge: Cambridge University Press.

———. 1901a. "G. F. FitzGerald." *Nature 63:*445–47. Repr. FitzGerald 1902, pp. xxii–xxix.

———. 1901b. "George Francis FitzGerald." *Physical Review 12:*292–313. Repr. FitzGerald 1902, pp. xxxix–lxiv.

———. 1914. "J. H. Poynting." *Phil. Mag. 27:*914–16. Repr. Poynting 1920, pp. xxiv–xxvi.

———. 1929. *Mathematical and Physical Papers.* 2 vols. Cambridge: Cambridge University Press.

Larmor, Joseph, ed. 1937. *Origins of Clerk Maxwell's Electric Ideas as Described in Familiar Letters to William Thomson.* Cambridge: Cambridge University Press.

Lee, George. 1950. "Oliver Heaviside: The Man." In Institution of Electrical Engineers, *Heaviside Centenary Volume,* pp. 10–17. London: Institution of Electrical Engineers Press.

Lodge, Oliver J. 1876a. "On a Mechanical Illustration of Thermoelectric Phenomena." *Phil. Mag. 2:*524–43.

———. 1876b. "On a Model Illustrating Mechanically the Passage of Electricity through Metals, Electrolytes, and Dielectrics, according to Maxwell's Theory." *Phil. Mag. 2:*353–74.

———. 1881. "The Relation between Electricity and Light." *Nature 23:*302–4. Repr. Lodge 1892b, pp. 369–84.

———. 1885. "The Identity of Energy: In connection with Mr. Poynting's Paper on the Transfer of Energy in an Electromagnetic Field; and on the Two Fundamental Forms of Energy." *Phil. Mag. 19:*482–87.

———. 1887. "Sketch of the Principal Electrical Papers Read before Section A during the Recent Meeting of the British Association at Manchester, 1887." *Electrical Review 21:*312–15.

———. 1888a. "Measurement of Electro-Magnetic Wave Length." *Elec. 21:*607–9.

———. 1888b. "On the Theory of Lightning Conductors." *Phil. Mag. 26:*217–30.

———. 1888c. "The Protection of Buildings from Lightning." *Elec. 21:*204–7, 234–36, 273–76, 302–3. Repr. Lodge 1892a, pp. 1–73.

———. 1888d. "The Protection of Buildings from Lightning." *Elec. 21:*303.

———. 1888e. "Sketch of the Electrical Papers in Section A at the Recent Bath Meeting of the British Association." *Elec. 21:*622–25, 660–63.

———. 1889a. "The Discharge of a Leyden Jar." *Elec. 22:*531–34. Repr. Lodge 1889b, pp. 359–83.

——. 1889b. *Modern Views of Electricity*. London: Macmillan.

——. 1890a. "The British Association Meeting at Leeds." *Elec. 25*:573–77.

——. 1890b. "Presidential Address to the Liverpool Physical Society." *Elec. 24*:579–80.

——. 1892a. *Lightning Conductors and Lightning Guards*. London: Whittaker.

——. 1892b. *Modern Views of Electricity*. 2d ed. London: Macmillan.

——. 1892c. "On the Present State of Our Knowledge of the Connection between Ether and Matter: An Historical Summary." *Nature 46*:164–65.

——. 1893a. "Aberration Problems." *Phil. Trans. 184A*:727–804.

——. 1893b. "Prof. Poynting's Still More Modern Views." *Elec. 31*:706.

——. 1894. "Heinrich Hertz." *Elec. 32*:273.

——. 1899. "Presidential Address." *Proc. Phys. Soc. 16*:343–86.

——. 1901. "George Francis FitzGerald." *Elec. 46*:701–2. Repr. FitzGerald 1902, pp. xix–xxii.

——. 1902. "George Francis FitzGerald, 1851–1901." *Year-Book of the Royal Society*, pp. 251–59. Repr. FitzGerald 1902, pp. xxxii–xxxix.

——. 1906. *Electrons*. London: George Bell.

——. 1909. *The Ether of Space*. London: Harper.

——. 1913. "Continuity." *BA Report*, pp. 3–42.

——. 1925a. "Oliver Heaviside: An Appreciation." *Elec. 94*:174–75.

——. 1925b. *Talks about Radio*. New York: George Doran.

——. 1931a. *Advancing Science, Being Personal Reminiscences of the British Association in the Nineteenth Century*. London: Ernest Benn.

——. 1931b. *Past Years: An Autobiography*. London: Hodder and Stoughton.

Lorentz, H. A. 1895. *Versuch einer Theorie der elektrischen und optischen Erscheinungen in bewegten Körpern*. Leiden: Brill.

——. 1909. *The Theory of Electrons*. Leipzig: Teubner. Repr. New York: Dover, 1952.

Lorenz, L. V. 1867. "On the Identity of the Vibrations of Light with Electrical Currents." *Phil. Mag. 34*:287–301.

McConnell, A. J. 1945. "The Dublin Mathematical School in the First Half of the Nineteenth Century." *Proceedings of the Royal Irish Academy 50A*:75–88.

MacCullagh, James. 1880. *The Collected Works of James MacCullagh*. Ed. Samuel Haughton and J. H. Jellett. Dublin: Hodges and Figgis, for Dublin University Press.

McDowell, R. B., and D. A. Webb. 1982. *Trinity College Dublin: An Academic History, 1592–1952*. Cambridge: Cambridge University Press.

Maxwell, James Clerk. 1873. *Treatise on Electricity and Magnetism*. 2 vols. Oxford: Clarendon.

——. 1881. *Treatise on Electricity and Magnetism*. 2 vols. 2d ed. Ed. W. D. Niven. Oxford: Clarendon.

——. 1890. *The Scientific Papers of James Clerk Maxwell*. 2 vols. Ed. W. D. Niven. Cambridge: Cambridge University Press. Repr. New York: Dover, 1952.

——. 1892. *Treatise on Electricity and Magnetism*. 2 vols. 3d ed. Ed. J. J. Thomson. Oxford: Clarendon. Repr. New York: Dover, 1954.

Maxwell, James Clerk, and Fleeming Jenkin. 1863. "On the Elementary Relations between Electrical Measurements." *BA Report*, pp. 130–63. Repr. Jenkin 1873, pp. 59–96.

Miller, Arthur, I. 1981. *Albert Einstein's Special Theory of Relativity: Emergence (1905) and Early Interpretation (1905–1911)*. Reading, Mass.: Addison-Wesley.

Mollan, Charles. 1981. "Science and Its Industrial Applications." In James Meenan and Desmond Clarke, eds., *The Royal Dublin Society, 1731–1981*, pp. 207–21. Dublin: Gill and Macmillan.

Nahin, Paul J. 1988. *Oliver Heaviside: Sage in Solitude*. New York: Institute of Electrical and Electronics Engineers.

O'Hara, J. G. 1975. "George Johnstone Stoney, F.R.S., and the Concept of the Electron." *Notes and Records of the Royal Society of London 29:*265–76.

O'Hara, J. G., and W. Pricha. 1987. *Hertz and the Maxwellians*. London: Peter Peregrinus.

Orr, William McFadden. 1900. "Considerations regarding the Theory of Electrons." *Phil. Mag. 50:*268–77.

Pais, Abraham. 1982. *Subtle Is the Lord: The Science and the Life of Albert Einstein*. Oxford: Oxford University Press.

Perry, John. 1899. "The Interdependence of Theory and Practice." *Elec. 43:*413–14.

Phillips, Melba. 1962. "Classical Electrodynamics." In S. Flügge, ed., *Encyclopedia of Physics 4:*1–108. Berlin: Springer.

Pihl, Mogens. 1972. "The Scientific Achievements of L. V. Lorentz." *Centaurus 17:*83–94.

Poynting, J. H. 1920. *Collected Scientific Papers*. Cambridge: Cambridge University Press.

Preece, W. H. 1885. "On the Relative Merits of Iron and Copper Wire for Telegraph Lines." *BA Report*, pp. 907–9.

——. 1887a. "Fast Speed Telegraphy." *Elec. 19:*423–26.

——. 1887b. "On Copper Wire." *Elec. 19:*372–73.

——. 1887c. "On the Coefficient of Self-Induction in Telegraph Wires." *Elec. 19:*400–401.

——. 1887d. "On the Limiting Distance of Speech by Telephone." *Proc. RS 42:*152–58. Repr. *Elec. 18:*395–97.

——. 1888. "Presidential Address to Section G." *BA Report*, pp. 781–92.

Preece, W. H., et al. 1888. "Discussion on Lightning Conductors." *Elec. 21:*644–48, 673–80.

Preston, Thomas. 1890. *The Theory of Light*. London: Macmillan. 2d ed. 1895. 3d ed., 1901. 4th ed., 1912.

Pupin, Michael I. 1924. *From Immigrant to Inventor*. New York: Scribner's.

Rayleigh [J. W. Strutt]. 1877. *The Theory of Sound*. 2 vols. London: Macmillan.

——. 1881. "On the Electromagnetic Theory of Light." *Phil. Mag. 12:*81–101.

Rayleigh [R. J. Strutt]. 1924. *Life of John William Strutt, Third Baron Rayleigh*. London: Edward Arnold. Repr. Ed. John N. Howard. Madison: University of Wisconsin Press, 1968.

——. 1943. *The Life of Sir J. J. Thomson*. Cambridge: Cambridge University Press.

Rowland, Henry. 1881. "On the Theory of Magnetic Attractions, and the Magnetic Rotation of Polarized Light." *Phil. Mag. 11:*254–61.

Rowlands, Peter. 1990. *Oliver Lodge and the Liverpool Physical Society*. Liverpool: Liverpool University Press.

Schaffner, K. F. 1972. *Nineteenth-Century Aether Theories*. Oxford: Pergamon.

Schilpp, P. A., ed. 1949. *Albert Einstein: Philosopher-Scientist*. 2 vols. Evanston, Ill.: Library of Living Philosophers.

Schuster, Arthur. 1910. "The Clerk-Maxwell Period." In *History of the Cavendish Laboratory, 1871–1910*, pp. 14–39. London: Longmans, Green.

Searle, G. F. C. 1896. "Problems in Electrical Convection." *Phil. Trans. 187A:*675–713.

——. 1950. "Oliver Heaviside: A Personal Sketch." In Institution of Electrical Engineers, *Heaviside Centenary Volume*, pp. 93–96. London: Institution of Electrical Engineers Press.

———. 1987. *Oliver Heaviside: The Man.* Ed. Ivor Catt. St. Albans: C.A.M. Publishing.

Siegel, Daniel. 1986. "The Origin of the Displacement Current." *HSPS 17:*99–146.

Siemens, Werner. 1966. *Inventor and Entrepreneur: Recollections of Werner von Siemens.* 2d ed. Trans. W. C. Coupland from 1st German ed., 1892. New York: Kelley.

Silliman, Robert H. 1963. "William Thomson: Smoke Rings and Nineteenth-Century Atomism." *Isis 54:*461–74.

Simpson, Thomas. 1966. "Maxwell and the Direct Experimental Test of His Electromagnetic Theory." *Isis 57:*411–32.

Smith, Crosbie, and M. Norton Wise. 1989. *Energy and Empire: A Biographical Study of Lord Kelvin.* Cambridge: Cambridge University Press.

Sommerfeld, Arnold. 1952. *Lectures in Theoretical Physics,* Vol. 3, *Electrodynamics.* Trans. E. G. Ramberg from 1st German ed., 1948. New York: Academic.

Spencer, J. B. 1970. "The Varieties of Nineteenth Century Magneto-Optical Discovery." *Isis 61:*34–51.

Stokes, G. G. 1862. "Report on Double Refraction." *BA Report,* pp. 253–82.

———. 1907. *Memoir and Scientific Correspondence.* 2 vols. Ed. Joseph Larmor. Cambridge: Cambridge University Press.

Stoney, G. J. 1885. "How Thought Presents Itself in Nature." *Proceedings of the Royal Institution 11:*178–96.

———. 1890a. "On Texture in Media, and on the Non-Existence of Density in the Elemental Ether." *Sci. Proc. RDS 6:*392–404.

———. 1890b. "Studies in Ontology, from the Standpoint of the Scientific Student of Nature." *Sci. Proc. RDS 6:*475–524.

———. 1897. "On the Proof of a Theorem in Wave-Motion." *Phil. Mag. 44:*98–102.

Sumpner, W. E. 1932. "The Work of Oliver Heaviside." *JIEE 71:*837–51.

Susskind, Charles. 1964. "Observations of Electromagnetic-Wave Radiation before Hertz." *Isis 55:*32–42.

Swenson, Loyd S. 1972. *The Ethereal Aether: A History of the Michelson-Morley-Miller Aether-Drift Experiments, 1880–1930.* Austin: University of Texas Press.

Tait, P. G. 1867. *Elementary Treatise on Quaternions.* Oxford: Clarendon.

Thompson, Silvanus P. 1881. *Elementary Lessons in Electricity and Magnetism.* London: Macmillan. 2d ed., 1895.

———. 1887. "Telephonic Investigations" [with discussion]. *JSTEE 16:*42–77, 81–105, 107–47.

———. 1901. "Presidential Address." *Proc. Phys. Soc. 17:*12–25.

———. 1910. *The Life of Sir William Thomson, Baron Kelvin of Largs.* 2 vols. London: Macmillan. Repr. as *Life of Lord Kelvin.* New York: Chelsea, 1976.

Thomson, J. J. 1880. "On Maxwell's Theory of Light." *Phil. Mag. 9:*284–91.

———. 1881. "On the Electric and Magnetic Effects Produced by the Motion of Electrified Bodies." *Phil. Mag. 11:*229–49.

———. 1883. *Treatise on the Motion of Vortex Rings.* London: Macmillan.

———. 1888. *Applications of Dynamics to Physics and Chemistry.* London: Macmillan.

———. 1931. "James Clerk Maxwell." In J. J. Thomson, ed., *James Clerk Maxwell: A Commemorative Volume, 1831–1931.* Cambridge: Cambridge University Press.

Thomson, William [Lord Kelvin]. 1882–1911. *Mathematical and Physical Papers.* 6 vols. Cambridge: Cambridge University Press.

———. 1887. "On the Vortex Theory of the Luminiferous Aether." *BA Report,* pp. 486–95.

——. 1888. "Sir William Thomson and Professor Clerk-Maxwell." Letter. *Times*, 17 Sept., p. 8. Repr. *Nature 38:*500.

——. 1889–94. *Popular Lectures and Addresses*. 3 vols. London: Macmillan.

——. 1904. *Baltimore Lectures on Molecular Dynamics and the Wave Theory of Light*. London: C. J. Clay.

Thomson, William, and P. G. Tait. 1867. *Treatise on Natural Philosophy*. Vol. 1. Oxford: Clarendon.

Topper, David R. 1971. "Commitment to Mechanism: J. J. Thomson, the Early Years." *AHES 7:*393–410.

Warwick, Andrew. 1989. "The Electrodynamics of Moving Bodies and the Principle of Relativity in British Physics, 1894–1919." Ph.D. diss., Cambridge University.

Wasserman, Neil H. 1985. *From Invention to Innovation: Long-Distance Telephone Transmission at the Turn of the Century*. Baltimore: Johns Hopkins University Press.

Weaire, Denis, and Seamus O'Connor. 1987. "Unfulfilled Renown: Thomas Preston (1860–1900) and the Anomalous Zeeman Effect." *Ann. Sci. 44:*617–44.

Whitehouse, Wildman. 1856. "The Law of Squares: Is It Applicable or Not to the Transmission of Signals in Submarine Circuits?" *Athenaeum*, 30 Aug., pp. 1092–93.

Whittaker, E. T. 1929. "Oliver Heaviside." *Bulletin of the Calcutta Mathematical Society 20:*199–220. Repr. in Chelsea edition of Heaviside 1893–1912, *1:*xiii–xxxiv.

——. 1951. *A History of the Theories of Aether and Electricity*, vol. 1, *The Classical Theories*. London: Nelson. Repr. New York: American Institute of Physics and San Francisco: Tomash, 1987.

Williams, L. Pearce. 1965. *Michael Faraday*. New York: Basic Books.

Wise, M. Norton. 1979. "William Thomson's Mathematical Route to Energy Conservation: A Case Study in the Role of Mathematics in Concept Formation." *HSPS 10:*49–83.

Index

Ingram Content Group UK Ltd.
Milton Keynes UK
UKHW040704230323
419035UK00004B/242